Robotic Sailing 2017

Kjell Ivar Øvergård
Editor

Robotic Sailing 2017

Proceedings of the 10th International Robotic Sailing Conference

Springer

Editor
Kjell Ivar Øvergård
Department of Maritime Operations
University College of Southeast Norway
Borre
Norway

ISBN 978-3-319-72738-7 ISBN 978-3-319-72739-4 (eBook)
https://doi.org/10.1007/978-3-319-72739-4

Library of Congress Control Number: 2017962996

Printed on acid-free paper

This Springer imprint is published by Springer Nature
The registered company is Springer International Publishing AG
The registered company address is: Gewerbestrasse 11, 6330 Cham, Switzerland

Preface

Autonomous surface vessels are currently a focus area for industry and academia. Efforts are now underway to produce MV Yara Birkeland, the first unmanned fully autonomous vessel for goods transport. The battery-powered vessel is an industry collaboration between Yara and Kongsberg Maritime and expected to replace 40.000 tuck-loads between the island Herya and ports in Brevik and Larvik. Significant resources are put into the development of this project. Similar initiatives are underway with both offshore supply vessels and small passenger ferries. Unfortunately, the same resources are not available for research and development of robotic sailing vessels, which has the advantage of using just clean, renewable and free energy for propulsion. Battery-powered vessels have the advantage that energy is readily available, but the production of batteries and electricity comes with a negative environmental footprint. The range of applications for zero-energy sailing vessels can be argued as limited compared to powered vessels, but sailing vessels have the advantages of being more environmentally friendly and allow for long sailing time. Utilizing wind for propulsion of vessels is a way to reduce emissions and return to an ecologically sustainable transport.

The 10th World Robotic Sailing Championship (WRSC) and the International Robotics Sailing Conference (IRSC) were organized by the University College of Southeast Norway (USN), with the support of Kongsberg Maritime, Norwegian Coastal Authorities, Norwegian Defence Research Establishment, and Horten municipality. Five vessels from USA, China, England, Åland and Norway participated in WRSC. Åland University of Applied Sciences won the class for larger vessels (between 1.5 and 4 m in length), and the small vessel class (for vessels smaller than 1.5 m) was won by the University of Southampton. For IRSC, seven papers reviewed by two or more reviewers were accepted and presented at the conference. All these papers are included in these conference proceedings. We would like to extend our thanks to the members of the program committee who was most helpful in reviewing the papers and giving helpful advice that greatly improved the papers contained herein.

In hindsight of the tenth IRSC, it is notable that the majority of papers focused similar challenges, but with different approaches. This is particularly interesting when we consider the scarceness of resources available for development and research in robotic sailing. The multitudes of approaches addressing the same problems seem to be a product of lack of coordination and coordination among the partners represented at IRSC. In a world with little resources and few partners involved, would it not be better to combine our resources and cooperate to solve the challenges to the betterment of all? One such research challenge that was apparent during the conference was the challenge of knowing the quality and suitability of different collision avoidance algorithms. Currently, in robotic sailing, most research papers only present one software or algorithm and then document the suitability of that software through simulations or field tests and there are no comparisons of different algorithms. One thing is to design and test algorithms for collision avoidance, but another thing is to find out which algorithm works best with respect to a number of criteria such as safety, efficiency, and adherence to the collision regulations. To do this, we must compare our newly developed algorithms with other state-of-the-art algorithms to see if our current solutions bring any improvements to the table. Unfortunately, such comparison of algorithms is not done, possibly because of the relative novelty of algorithms for autonomous navigation or the lack of a standard for how to compare different algorithms. Hence, there is a need for a unified standard for testing and evaluating different algorithms for route selection, navigation, and collision avoidance. Hence, we need a standardized and structured way of testing algorithms and software for autonomous sailing. This requires several novel steps: First, there is a need to determine the criteria that defines navigational quality, generally and specifically for sailing vessels, and second, we must create a testing environment that allows for the testing of different algorithms under identical circumstances. With this aim in mind, we hope that future IRSC will take on this challenge to define a standard for assessment of algorithms for robotic sailing and to compare different solutions to collision avoidance, route selection, surveying, and navigation in a rigorous way.

We would like to thank the organizations that supported WRSC and IRSC2017 financially: Norwegian Coastal Administration, Kongsberg Maritime, HSN Industry Academy, Norwegian Defence Research Establishment, and Horten municipality. The WRSC and IRSC2017 would not be possible without the effort of the organizing committee. We extend our gratitude to everyone involved with the committee: Inger Norunn Nergaard, Per Haavardtun, Anders Kristiansen, Geir Varholm, Cathrine Eyde, and Ragnar Niemi Berg.

Horten, Norway
August 2017

Kjell Ivar Øvergård
Marius Stian Tannum

Organizing Committees

General Chair

Marius Stian Tannum, University College of Southeast Norway, Borre, Norway

IRSC Program Chair

Kjell Ivar Øvergård, University College of Southeast Norway, Borre, Norway

Program Committee

José Carlos Alves, Universidade do Porto/INESC TEC, Portugal
Fabrice Le Bars, École Nationale Supérieure de Techniques Avancées (ENSTA) Bretagne, France
Ole Blaurock, Fachhochschule Lübeck, Germany
Jan Kåre Bording, University College of Southeast Norway, Borre, Norway
Noureddine Bouhmala, University College of Southeast Norway, Borre, Norway
Benoit Clement, École Nationale Supérieure de Techniques Avancées (ENSTA) Bretagne, France
Nuno A. Cruz, Universidade do Porto/INESC TEC, Portugal
Bjørn-Olav Holtung Eriksen, Norwegian University of Science and Technology, Trondheim, Norway
Andreas Flaten, Norwegian University of Science and Technology, Trondheim, Norway
Anna Friebe, Å land University of Applied Sciences, Finland
Oren Gal, Technion Israel Institute of Technology, Israel
Jorge Cabrera Gamez, Universidad de Las Palmas de Gran Canaria, Spain
Erik Maehle, University of Lübeck, Germany
Benedita Malheiro, Instituto Superior de Engenharia do Porto, Portugal
Paul Miller, US Naval Academy, USA
Cedric Pradalier, GeorgiaTech Lorraine, France

Kostia Roncin, École Nationale Supérieure de Techniques Avancées (ENSTA) Bretagne, France
Colin Sauze, Aberystwyth University, Wales, UK
Michael Schukat, National University of Ireland, Galway, Ireland
Alexander Schlaefer, Hamburg University of Technology, Germany
Halvor Schøyen, University College of Southeast Norway, Borre, Norway
Diedrich Wolter, University of Bremen, Germany

Contents

Part I
Sails and Software for Autonomous Sailing

An Alternative Wing Sail Concept for Small Autonomous Sailing Craft

Paul Miller, Carolyn Judge, Dylan Sewell and Sean Williamson

Abstract An alternative wing sail section shape was developed as a potential improvement to either a soft sail or the common NACA00XX section used on some small autonomous sailing craft. The prototype wing was developed as an evolution of a traditional square sail. Evaluation included wind tunnel and on-the-water tests followed by performance estimates from a Velocity Prediction Program. The test vessel was the 1.2 m long Version 2 MaxiMOOP. Advantages of the wing sail over the current voyaging soft sail design include: higher speeds in some conditions, increased durability, increased buoyancy in the case of capsizing, and easier implementation of control systems. Some potential theoretical advantage was found over the NACA0009.

1 Introduction

The search for speed under sail is never ending. Vessel designers constantly develop sailboat designs for improved performance and reliability. With the hopes of enhancing future SailBot designs, this paper explores an alternative wing sail concept for small autonomous (computer-controlled) vessels. The test platform was a 1.2 m long MaxiMOOP design (Version 2), developed by Aberystwyth University and the United States Naval Academy in 2012 [12]. To-date, over 25 MaxiMOOPs have been built and one, ABoat Time, has sailed over a thousand ocean miles. The alternative concept features a permanently cambered airfoil section that is symmetrical about the 50% chord line, rather than about the midplane as is common with airfoils such as

P. Miller (✉) · C. Judge · D. Sewell · S. Williamson
US Naval Academy, Annapolis, MD, USA
e-mail: phmiller.pe@gmail.com

C. Judge
e-mail: judge@usna.edu

D. Sewell
e-mail: Dylan.W.Sewell@gmail.com

S. Williamson
e-mail: snwilliamson02@gmail.com

K. I. Øvergård (ed.), *Robotic Sailing 2017*,
https://doi.org/10.1007/978-3-319-72739-4_1

the NACA00XX, where the XX represents the thickness as a percent of chord length (e.g. NACA0012). This means trimming the wing is done in a manner similar to a traditional square sail rather than like a modern airfoil.

Wing sails have featured in numerous sailboat designs. One good reference is C.A. Marchaj's epic 1979 book, Aero-Hydrodynamics of Sailing, [10] which includes discussion of wing sails as far back as 1941. Those craft primarily used single-element NACA00XX foils with a trim tab to control the wings angle of attack. Unfortunately, as he noted on pages 120–121, the maximum lift of the symmetrical NACA00XX wing (3D C_L versus 2D section C_l), which drives light wind (low Reynolds Number) performance is limited to about 0.8, or about half of what a soft sail could produce. A highly cambered, or multiple element foil however, could see significantly higher lift coefficients as illustrated on page 312 where a thick, cambered airfoil produced a C_L of 1.32.

Multi-element wing sails are now commonly used in the America's Cup, although their development began in the C-Class catamarans for the Little America's Cup. Their design criteria significantly differ from those for small autonomous vessels due to their narrower apparent wind angle range and much lower durability requirements.

Rigid wing sails for autonomous vessels include the Atlantis Project at Stanford in 2001 [3], Harbor Wing X-1 in 2006 [6], multiple boats from Aberystwyth University in 2009–2011 [13, 14], Saildrone in 2012 [7], A-TIRMA G2 in 2015 [2] and Cornell's boat in 2016 [1]. All of these used symmetrical, single element foils. Cambered single element wing sails are much less common, although the DynaRig, consisting of folding, thin sails were successfully used on the large yacht, Maltese Falcon, launched in 1990 [8]. With her masts located on the wings windward side, undesirable flow disruption is created.

2 Design Development

Inspiration for this alternative wing sail originated from a paper written by former United States Naval Academy professor, Thomas Gillmer and a sail by one of the authors on a replica vessel in Roskilde, Denmark. Professor Gillmer wrote extensively about ancient vessels, including ships used by the Vikings [5]. An example of the square sail to which Professor Gillmer referred is shown in Fig. 1. In particular, he commented about the inherent practicality of the square sail and how modern improvements could be made to that rig and sail to increase their efficiency. Professor Gillmer postulated that a second sail placed on the windward side of the mast, with the two sails vertical edges sewn together, would reduce the inefficiencies introduced by the mast and create a better airfoil shape. The advantages of the square sail design are that the sail has a large area to catch wind on a run and it pivots about the center of the sail so it is easier to trim [5]. While sailing a Viking ship replica at Roskilde, Professor Miller noted that although the rig allowed the vessel to point high, its progress to windward was impeded by significant leeway caused by little lateral keel area.

Fig. 1 Square sail configuration—Norse traveling ship A.D. c 1000 [5]

This project sought to improve the Viking square sail by creating a similarly shaped foil design using a rigid wing that changed the straight yard (spar) to a cambered shape and increased the aspect ratio. The wing thickness was set as the mast thickness for the free-standing spar. The key dimensions of the wing sail were based off previous designs of the MaxiMOOPs voyaging soft sail. These included the same span, aspect ratio and total projected area. Previous tests with soft sails on the MaxiMOOP V.2 found that an aspect ratio near three generated the best performance over different winds and headings [12]. Figure 2 shows the soft sail and prototype wing.

As the project's goal was to produce a sail that evolved the square sail concept the wing's characteristics included symmetry about the 50% chord line, sharp leading/trailing edges, and camber. The maximum curvature gradient occurs near the mast with the leading/trailing edges fairly straight. Section thickness was driven by mast bending strength, which for the MaxiMOOP resulted in a 16 mm diameter mast. Camber was based on typical soft-sail designs, with the bottom camber set at 6, 12% in the middle and 10% at the top (in the third wing version it was changed to 6/10/12). The larger camber at the top is hypothesized to produce a more forgiving airflow in response to the greater angle of attack variation caused by the vessel rolling. To reduce induced drag and heeling moment the planform is tapered. Total projected area is 0.3 m^2. Figure 3 compares the wing section to a NACA0009 and Fig. 4 presents the Table of Offsets.

Weight aloft is critical to vessel performance as stability counteracts the sail forces and the amount of thrust produced. To reduce weight the wing was NC-milled from 48 kg/m^3 Dow STYROFOAM Insulation board. The 16 mm diameter carbon fiber mast extended 60% up the sail. Produced by Rock West Composites it featured 80% uni-directional fibers, 20% hoop fibers and had a wall thickness of 1.6 mm. The wing

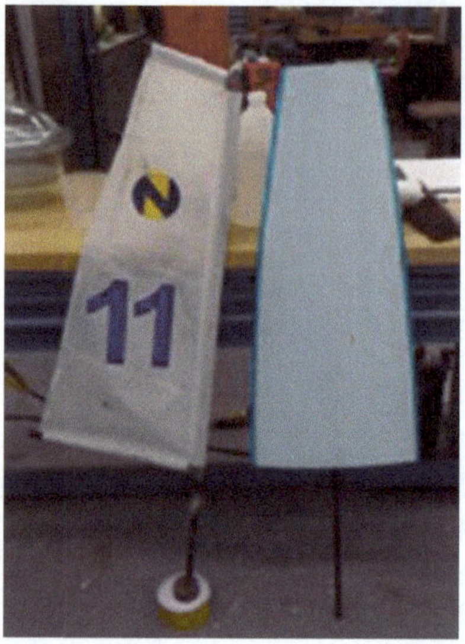

Fig. 2 Comparison of the current voyaging soft sail and the new wing sail design

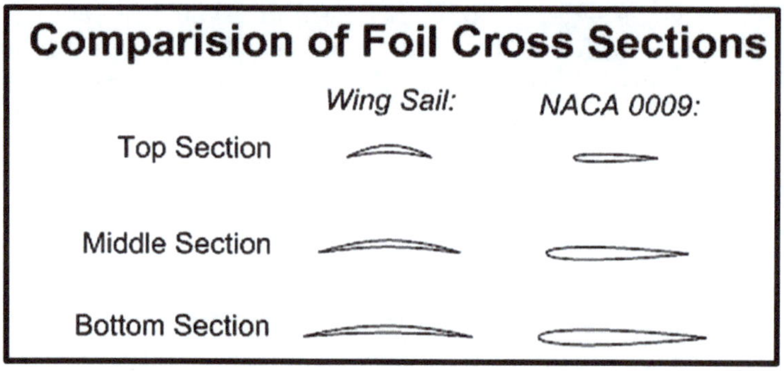

Fig. 3 Section shapes of the wing sail design and a similarly sized NACA0009

was covered by one layer on each side of 80 g/m^2 E-glass cloth set in WEST 105/206 epoxy. The finished weight of the wing sail was exactly the same as the soft sail, 0.68 kgf.

Although the wing was designed for a Version 3 MaxiMOOP (see Fig. 5) the on the water testing was performed on the USNA test platform, a Version 2 MaxiMOOP. Both vessels are described in reference [12].

Offsets for MaxiMOOP wing (mm)					
Bottom			Top		
X (chord)	Y fwd face	Y aft face	X (chord)	Y fwd face	Y aft face
0	26	26	100.5	28	28
18.3	21.7	24.3	109.6	23.2	25.9
36.5	17.7	22.8	127.9	15.2	22.5
54.8	14.1	21.4	146.2	8.8	19.8
73.1	10.9	20.2	164.5	4	17.7
91.4	8.1	19.1	182.7	1	16.4
109.6	5.6	18.2	201	0	16
127.9	3.6	17.4	219.3	1	16.4
146.2	2.1	16.8	237.5	4	17.7
164.5	0.9	16.4	255.8	8.8	19.8
182.7	0.2	16.1	274.1	15.2	22.5
201.0	0	16	292.4	23.2	25.9
219.2	0.2	16.1	301.5	28	28
237.5	0.9	16.4			
255.8	2.1	16.8			
274.1	3.6	17.4			
292.3	5.6	18.2			
310.6	8.1	19.1			
328.9	10.9	20.2			
347.2	14.1	21.4			
365.4	17.7	22.8			
384.1	21.7	24.3			
402.8	26	26			

Fig. 4 Offsets for alternative wing

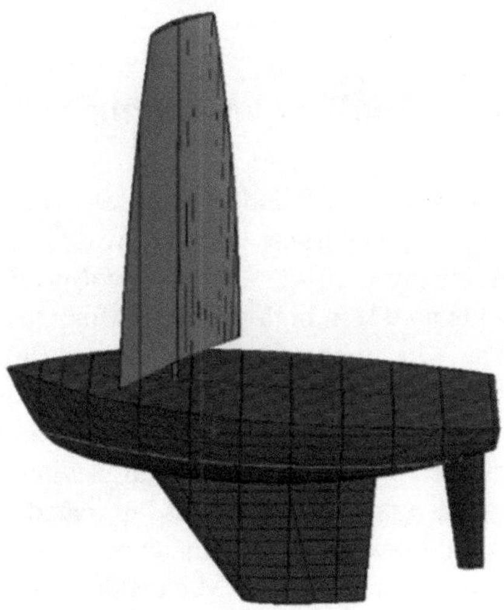

Fig. 5 MaxiMOOP Version 3 with wing sail MaxSurf model

3 Evaluation Approach

Three different methods were attempted to evaluate the wing sail—computational fluid dynamics (CFD), wind tunnel experiments, and on-the-water tests. Used together, these tests would hopefully provide strong evidence to support the use of the wing sail for future SailBot designs. The goal was to quantify the lift, drag, and yaw moment. The first two would be used in a Velocity Prediction Program (VPP) to estimate the vessel's performance, which could then be compared to the soft sail and the yaw moment would be used to size the actuators.

3.1 Computational Fluid Dynamics

In the hopes of having a versatile method for analyzing different configurations of sail and/or wing shape, the wing sail was investigated using Computational Fluid Dynamics (CFD). Evaluation of 2D sections of the wing was attempted using AutoDesk CFD 2016 and XFOIL. The full 3D sail was considered using ANSYS Fluent. Unfortunately, issues such as the shape of the wing (specifically the sharp leading and trailing edges), meshing difficulties, the low Reynolds' Number involved, licensing issues, and time constraints resulted in a failure to conclude any of the CFD analyses.

3.2 Wind Tunnel Testing

Although the lift and drag performance of the wing sail could not be evaluated using CFD, a comparable wing sail was tested in a low speed wind tunnel in the US Naval Academy's Closed Circuit Low-Speed Wind Tunnel. The wind tunnel has a 107 cm wide by 97 cm high by 244 cm long test section which allowed for full scale testing. The wind tunnel is powered by a 186 KW motor and reaches a maximum speed of approximately 67 m/s. The wind tunnel is equipped with an external Schenck platform force balance, with the capability to measure lift, drag, side force, pitching moment, rolling moment, and yawing moment.

The wing sail tested was fabricated using a hot wire cutter. As with the prototype it was built with Styrofoam covered with fiberglass. It differed slightly from the actual wing sail used on the MaxiMOOP as it had a stainless steel mast (compared with a carbon mast), the elliptical planform had a straight taper, and the projected area of the sail was 0.24 m^2 (compared with 0.3 m^2). Figure 6 shows a picture of the wing sail during testing. The blockage varied from less than 1% at zero angle of attack (the sail aligned with the flow direction) to 18% at an angle of attack of 90°. The wind speed in the test section was 10 knots (5.1 m/s), resulting in Reynolds' Numbers ranging from 70,000 to 100,000, depending on the chord. The sail was tested at angles of attack ranging from 0° to 180°. A force balance connected to the mast measured forces on the sail at each angle of attack, allowing calculations of the lift and drag coefficients. No twist was observed. The C_D and C_L results are shown in Fig. 7.

The lift coefficient data shown in Fig. 7 shows the lift increasing with angle of attack until reaching a maximum value of 1.10 at an angle of 12°, after which the sail shows stall. As the angle of attack continues to increase, the lift coefficient decreases until reaching a minimum lift of near zero when the sail is oriented at 80° (almost perpendicular to the flow). In this configuration, the sail is acting as the square sail inspired by it and the drag coefficient for this condition is at its maximum value of 1.71. As the sail continues to rotate past 80°, the lift coefficient begins to increase again, reaching a secondary maximum of 1.01 at an angle of 168°. The drag coefficient is at a minimum when the sail is aligned with the air flow and maximized when the sail is close to perpendicular to the flow.

Figure 8 compares the wind tunnel results for the wing sail with published lift and drag coefficients for a NACA 0009 foil tested at a Reynolds Number of 100,300. The C_L and C_D curves for the two shapes are very similar, although in agreement with Marchaj's findings the cambered foil's lift coefficients are a little higher. However, the NACA 0009 foil stalls at an angle of attack close to 9°, while the wing sail doesn't stall until about 12°. The drag coefficient is very similar for these shapes for the angles of attack there is data available for the NACA 0009 foil. At very high angles of attack a cambered foil typically has a higher drag coefficient than a symmetrical foil.

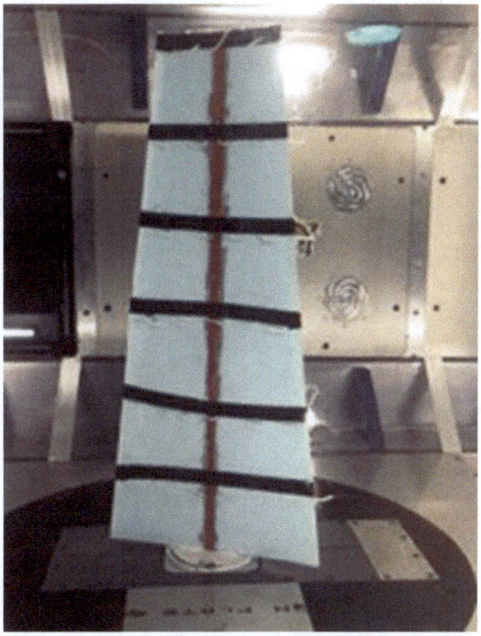

Fig. 6 Wing sail wind tunnel test conducted in USNAs low speed wind tunnel at 10 knots and an angle of 20°

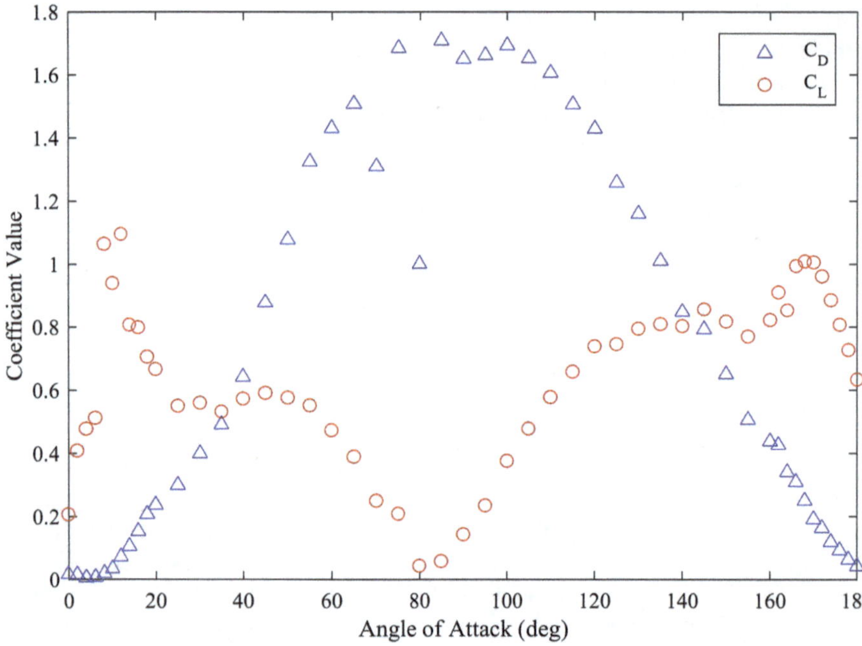

Fig. 7 C_L and C_D at different angles of attack for the wing sail conducted in USNAs low speed wind tunnel

The drag measured in the wind tunnel includes the induced drag, C_{Di} (drag due to the lift generated). The two-dimensional section drag coefficient is found by subtracting the induced drag from the drag coefficient calculated from the wind tunnel measurements,

$$C_d = C_D - C_{Di} \tag{1}$$

where

$$C_{Di} = \frac{C_L^2}{\pi A} \tag{2}$$

and $A = (\text{span})^2/\text{area}$. The two-dimensional sectional lift coefficient is considered equal to the three-dimensional lift coefficient calculated from the wind tunnel measurements, except at the maximum values. At the maximum lift coefficient values the spanwise lift distribution can result in a reduction of lift as large as 15% [15]. Therefore, the 2D sectional lift coefficients were determined by dividing the lift coefficient calculated from the wind tunnel measurements by 0.85,

$$C_l = \frac{C_L}{0.85} \tag{3}$$

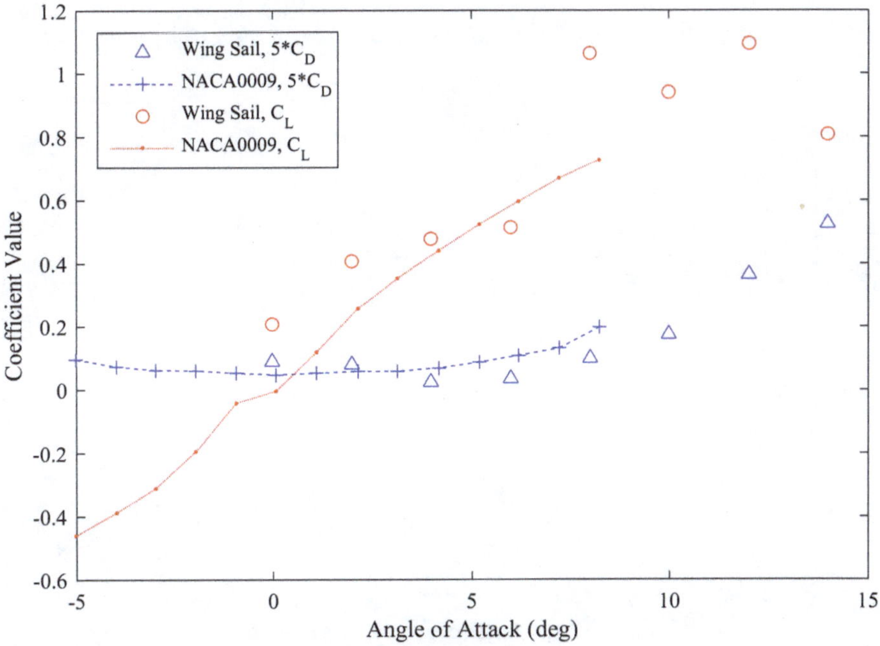

Fig. 8 C_L and C_D at different angles of attack for the wing sail compared to NACA0009 at Reynolds number 100300 [9]

The corrected 2D lift and drag coefficients are shown in Fig. 9. The trends are the same as seen in Fig. 7, but there are slight differences in values and in the shape of the drag coefficient curve. The corrected maximum lift coefficient still occurs at 12°, but the value is 1.29 (an increase from the C_L value of 1.1). As would be expected, the biggest changes in the drag coefficient occur for the conditions where the lift coefficients is the largest.

An important measurement taken during the wind tunnel experiment was the yaw moment due to changes in the center of pressure acting on the sail. Figure 10 shows the yaw moment versus angle of attack. The yaw moment is maximum for conditions when the sail is oriented perpendicular to the flow direction. In general, the yawing moment decreases as the angle of attack approaches zero or 180°, but the data is not consistent on this. The forces measured were very small leading to relatively large uncertainties in the moment measurements. However, the maximum moment magnitude can be scaled so as to determine the proper actuator. The maximum value measured during the wind tunnel testing was 0.29 N-m at an angle of attack of 85°.

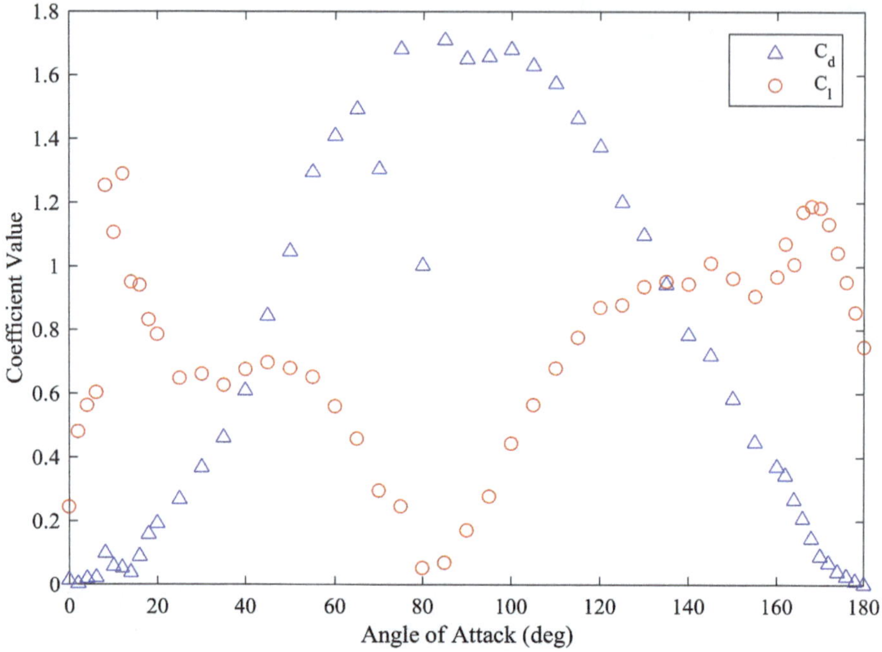

Fig. 9 C_l and C_d at different angles of attack for the wing sail conducted in USNA's low speed wind tunnel

3.3 Velocity Prediction Program (VPP)

VPPs are widely used to predict sailboat performance. Their basic concept solves the static equilibrium conditions of surge, roll, and sway beginning with a wind speed and direction to establish the forces and moments produced by the sails. Then using the vessel's characteristics to calculate stability and resistance, the code iterates to an equilibrium condition. A more detailed description is available in Principles of Yacht Design [9]. For our evaluation we used an Excel-based spreadsheet named PCSAIL (version 2.53) developed by David Martin and Robert Beck from the University of Michigan [11] that uses the Delft series of models. Users may select either Claughton or Larsson sail coefficients or may input their own. C_l and C_d from our wind tunnel results were used for the wing. Figure 11 shows the difference between the soft and wing sails in winds from 11 to 30 knots.

The VPP suggests that the wing sail does not perform as well as the soft sail when going close hauled up to a wind angle near a close reach. From there the two sails are about even until sailing a broad reach where the wing becomes increasingly faster the deeper the boat sails. Experience from the soft sail has shown that the boat does not perform well pointing high in chop. To address this a practical limit of 50° true wind angle is used. With the wing at that true wind angle, the boat will see a potential a 0.1–0.2 knot speed decrease.

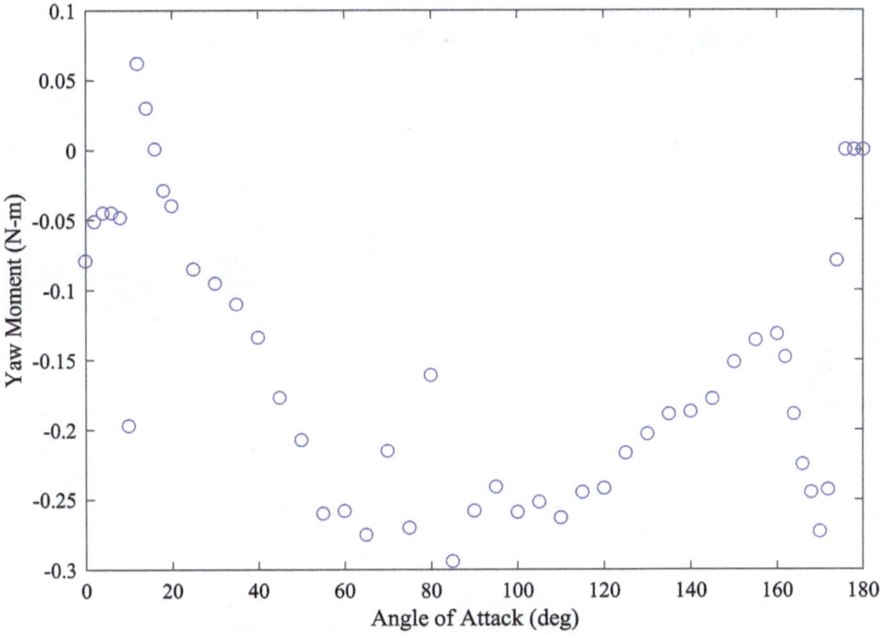

Fig. 10 Yaw moment versus angle of attack

Using the oceanic routing method described in [4] for a 3000 NM voyage from Cape Cod to Ireland for the microTransAtlantic Challenge, where although roughly 85% of the course is downwind, the loss of upwind speed is still important, with the predicted time for the wing at 62 days, versus 58 for the soft sail, a loss of 4 days (7%) in transit time.

An important caveat in believing the VPP however is that the VPP assumes the soft sail has a maximum C_l of 1.7, which assumes a triangular sail plan. Although the VPP adjusts C_L and C_D for aspect ratio, as the MaxiMOOP has a rotating, fixed gaff and boom sail it is likely its max C_l is lower than assumed by the VPP. Indeed, the sea trials indicated different results. Additionally the VPP does not calculate yaw moment (weather helm), which adds drag, but is much reduced for the wing sail as it is balanced over the hull while the soft sail is mostly on one side, creating more weather helm.

3.4 Sea Trials

Sailing tests were run on the Severn River with the wing sail to test its application in real-world settings and compare to our VPP and wind tunnel testing. The tests were conducted in winds greater than 5 knots in order to collect reliable data. Winds

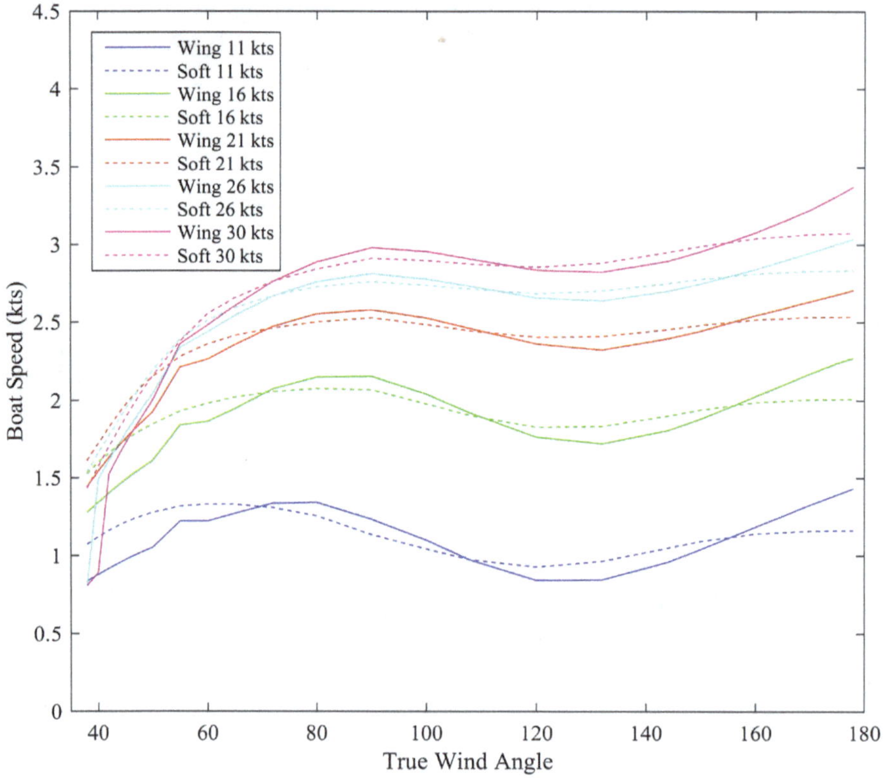

Fig. 11 VPP results: comparison of predicted boat speed at different wind angels for soft sail and wing sail

ranged from 7 to 13 knots over the course of the trials. Using a GPS placed inside the hull of the MaxiMOOP, time, position, course, and overall speed were recorded. Wind speeds and directions were determined using an anemometer held in the RHIB at about the same height off the water as the wing. Data is shown in Fig. 12.

Figure 12 shows the vessel's speed over the course of a trial run. Data was taken at 1 Hz and the lines represent continuous data points. For comparison, the horizontal dotted line shows the maximum speed recorded by a MaxiMOOP with the voyaging soft sails in any wind condition during all of its offshore sailing [12] and represents a speed slightly higher than "hull speed". Trial run times varied due to vessel traffic on the Severn River as well as operator error when attempting to keep the wing sail consistently trimmed. The longest data stream collected was going close-hauled in 13 knots. The average speed in those conditions was 1.5 knots, while the VPP predicted 1.3; the theory under predicting actual performance by 15.

That the wing sail was able to provide enough thrust to generate speeds above the 2.7 knot benchmark at multiple points of sail suggests that the wing sail is capable for use in future SailBots. Of note, these speeds were achieved in as little as 7–13

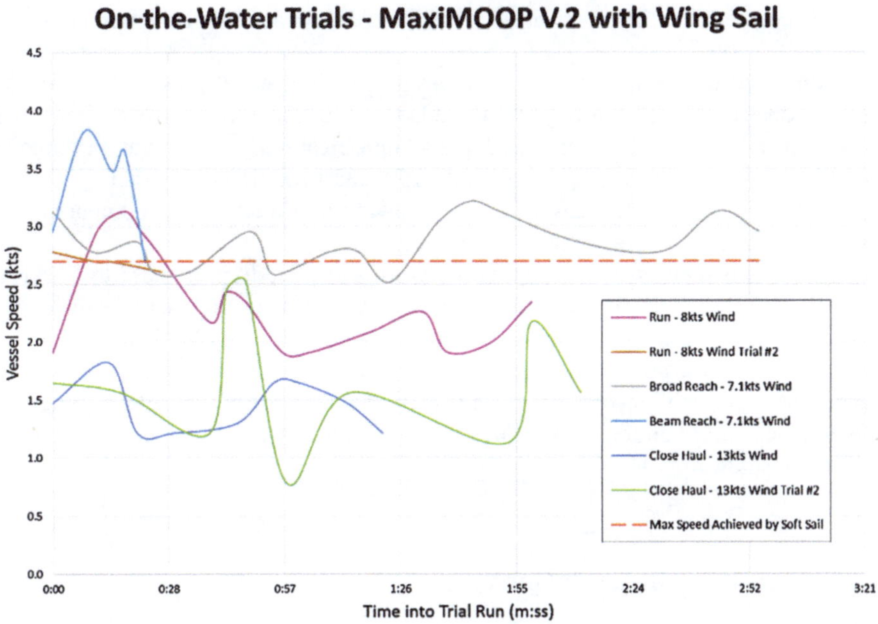

Fig. 12 Sea trial runs at different points of sail for MaxiMOOP showing vessel speed over the course of a trial run

knots of wind. While the wing sail is designed to be able to sail in 30 knots of wind the speeds observed may suggest that the wing sail could overpower the vessel in higher winds and that a smaller sized wing sail would be more appropriate. The VPP predicted over 35° heel while beam reaching in 30 knots.

Unfortunately, good underway data for the soft sail is lacking, but as a qualitative test the two rigs were raced against each other. The new MaxiMOOP Version 3 was sailed with the soft sail rig while the Version 2 boat had the prototype wing. The winds were quite light, ranging from calm to four knots, however the trend seemed to indicate that the two boats were about the same speed upwind and reaching while the wing was clearly faster downwind.

4 Discussion

The wind tunnel and on the water test results suggest that the new wing sail design concept will provide equivalent or perhaps improved performance compared to the soft sail. On-the-water tests also suggest however that the VPP results may overestimate the soft sail performance and/or underestimate the wing sail's.

In addition to the inherently different C_l and C_d for a triangular versus a quadrilateral sail, the wing may provide greater performance simply because its shape is fixed. The sail cloth stiffness causes the sail shape to distort until the wind reaches about 8–10 knots.

Potential advantages of the wing sail design also include increased durability as the wing does not "luff" in heavy winds, added buoyancy in the case of capsizing and consistent sail shape. Also, compared to a symmetrical wing, the actuation range is significantly less, requiring approximately 150° mast rotation compared to 330 for a normal wing. Actuation loads are small, allowing for direct control compared to a trim tab that increases drag and weight up high.

Some disadvantages of the alternative wing sail include proneness to backing because the wing sail does not luff, difficulty in analyzing the sharp leading and trailing edges of the wing sail in CFD (which limits development), and possibility of the vessel being overpowered in high wind conditions. It is conceivable that the wing can be feathered for high wind, downwind, sailing. As the wing is rigid, transportation to the launch site might be more difficult, although a vertically segmented design is a potential solution.

5 Future Recommendations

There are some areas of uncertainty that are worth exploring in future testing. CFD will provide a powerful tool for comparing wing shapes and optimizing wing design. It is worth working through the difficulties encountered so as to benefit future wing sail development for autonomous sailing vessels. Suggestions include creating models with more rounded leading and trailing edges, identifying the best software for handling low Reynolds' Number flow, and exploring the best meshing methods for this type of uniquely shaped foil. More work using wind tunnel analysis would be beneficial by exploring increased wind speeds and considering various section shapes, including a NACA0009 with the same area, span and planform. A wind tunnel test on the voyaging soft sail would be of great benefit to determine its actual lift and drag coefficients for aid in evaluating the performance improvements of the wing sail. For the VPP analysis, one relatively easy comparison would be to analyze various symmetrical sections for comparison to the soft sail and alternative wing. Lastly, more extensive on-the-water tests with heavy winds and sea conditions would allow observation into the effects on the wing sail in harsh conditions. Ultimately the wing will also require modification to the control code with the logic to recover from backwinding.

Acknowledgements We are very grateful for the help of the staff working in the US Naval Academy's Technical Services Department, and in particular the Model Shop (Dave Majerowicz, Tom Price and Dale Boyer) and the Aerodynamics Laboratory (Louise Bechnel, Daniel Rodgerson, George Willman, and Professor Joseph Milluzzo).

References

1. Augenstein, Thomas, et al. 2016. Using a controlled sail and tail to steer an autonomous sailboat. In *9th IRSC*, Portugal.
2. Donimhuez-Brito, Antonio C., et al. 2015. A-TIRMA G2: An oceanic autonomous sailboat. In *8th IRSC*, Finland.
3. Elkaim, G.H. 2001. System identification for precision control of a wingsailed GPS-guided catamaran. PhD thesis, Stanford University, CA.
4. Gibbons-Neff, Peter, and Miller Paul. 2011. Route planning for a micro-transat voyage. In *4th IRSC*, Germany.
5. Gillmer, Thomas Charles. 1972. Working watercraft and the ancient boat. unpublished manuscript.
6. http://harborwingtech.com/products_hwt_x-1.htm . Accessed 16 May 2017.
7. http://saildrone.com/#Technology . Accessed 16 May 2017.
8. https://www.symaltesefalcom.com/about.php . Accessed 16 May 2017.
9. Larsson, L., R. Eliasson, and M. Orych. 2014. Principles of yacht design, 4th edn. Camden, Maine: McGraw Hill.
10. Marchaj, Czeslaw A. 2000. *Aero-hydrodynamics of sailing*. London: Aldard Coles Nautical.
11. Martin, David E., and Robert F. Beck. 2001. PCSAIL, A velocity prediction program for a home computer. In *15th Chesapeake sailing yacht symposium*, 99–116.
12. Miller, P., et al. 2015. MaxiMOOP: a multi-role, low cost and small sailing robot platform. In *Proceedings of the 7th IRSC*, Ireland, 2014.
13. Neal M., C. Sauze, and B. Thomas. 2009. Technologies for autonomous sailing wings and wind sensors. In *2nd IRSC*, Portugal.
14. Sauze C., and M. Neal. 2011. MOOP: A miniature sailing robot platform. In *4th IRSC*, Germany.
15. Selig, M., John F. Donovan, and David B. Fraser. 1989. *Airfoils at low speeds*. Virginia, USA, H.A: Stokeley Publisher.

Correlation of Velocity Prediction Program for Small Autonomous Sailboats

Paul Miller, Peter Hodapp, Ryan Pinch, Jessica Robinson and Camille Tiemann

Abstract Velocity Prediction Programs (VPP) are used to improve sailboat design and estimate performance. Using the concept of static equilibrium they rely on empirical relations derived from tank testing of a series of standard yacht hull shapes to estimate resistance. The VPP accuracy for non-standard hull shapes, such as SailBots, is a concern. Correlation of the VPP estimates to on-the-water measurements of a small autonomous sailboat was this project's goal. Sixty-two data points were taken in winds ranging from 5 to 10.6 knots using the United States Naval Academy (USNA) SailBot, *Mid-lenium FalcoN** and were compared to PCSail2.53, an Excel-based VPP developed from the Delft series and standard VPP code. From the analysis, it is concluded that the VPP makes an acceptably accurate prediction for boat speed within an average wind speed between 5 and 7 knots. At higher winds speeds, some points were within 95% confidence; however, the majority of on the water values were significantly slower than the VPP predictions.

1 Introduction

Velocity Prediction Programs (VPPs) are widely used to develop sailboat designs as they allow rapid feedback of design changes on performance. VPPs predict boat speed, heel angle, leeway and other performance values for a given wind speed at varying wind angles by balancing the forces and moments the boat experiences. For

P. Miller (✉) · P. Hodapp · R. Pinch · J. Robinson · C. Tiemann
United States Naval Academy, Annapolis, MD, USA
e-mail: phmiller.pe@gmail.com

P. Hodapp
e-mail: m182670@usna.edu

R. Pinch
e-mail: m185148@usna.edu

J. Robinson
e-mail: m185520@usna.edu

C. Tiemann
e-mail: m186492@usna.edu

K. I. Øvergård (ed.), *Robotic Sailing 2017*,
https://doi.org/10.1007/978-3-319-72739-4_2

example, in surge, the thrust produced by the sails is determined from the standard lift and drag equations where the coefficients of lift and drag were determined by wind tunnel and full-scale testing. Opposing the sail thrust is the hull resistance. An empirical formula for resistance uses the hull, keel and rudder characteristics to determine drag coefficients. The program then iterates the resistance formula by varying speed until thrust equals resistance.

VPPs are generally thought to indicate the impact of design trends, but their absolute accuracy for a particular boat type depends on the limitations of the empirical formulas used to develop the hull resistance and aerodynamic force estimates [2]. Apart from event-specific boat types, such as those for the America's Cup or round-the-world races, most VPPs, including the PCSail spreadsheet used in this project, use the Delft series of tank test model data that are based on standard hull shapes for vessels of 10–25 m in length and are generally considered accurate to within 6–9 sec/NM for typical moderate displacement sailboats [3]. PCSail was used for this project as it is no-cost, runs easily and quickly on Excel, has no hidden equations and is generally robust enough to run small craft. No literature was identified that provided correlation for boats under seven meters length. VPPs were used in the development of all fourteen USNA autonomous sailboats built to-date in spite of not having data that confirmed their accuracy. In particular, the VPP was used to estimate the voyage time for the MicroTransAtlantic Challenge boats [1]. Significant errors would change the optimized routing results. Gaining confidence in the VPP estimates was therefore the main goal of this project.

To conduct this study the autonomous two-meter long SailBot, *Mid-lenium FalcoN** (MF),[1] was sailed under remote control (Fig. 1). This was chosen over autonomous control as the sails could be trimmed to greater accuracy and the point of sail was better controlled. The boat was sailed on a constant course with the wind direction, wind speed and boat speed recorded. From the time series an average boat speed was determined that was then correlated to the VPP estimate (Fig. 2).

MF is a double-chine plywood boat evolved from the single-chine WRSC winning boat, *SeaQuester*. MF's principal dimensions are shown in Table 1 and the sail dimensions in Table 2. Note that the characteristics are in United States Customary units due to the spreadsheet's limitations. Copies of the freeware PCSail are available from the authors.

2 Approach

The purpose of this research was to compare measured to predicted boat speeds and the main challenge was acquiring acceptable data. The VPP, PCSail2.53, uses the sail dimensions and hull parameters as inputs to predict the boat speed at different wind speeds and angles. Although documenting the boat's dimensions was straight

[1]The asterisk after the capital N in MFs name comes from a Naval Academy tradition of adding a star after the Navys N to signal victory over an Army team.

Fig. 1 USNA SailBot *Mid-lenium FalcoN** sailing in six knots during data acquisition

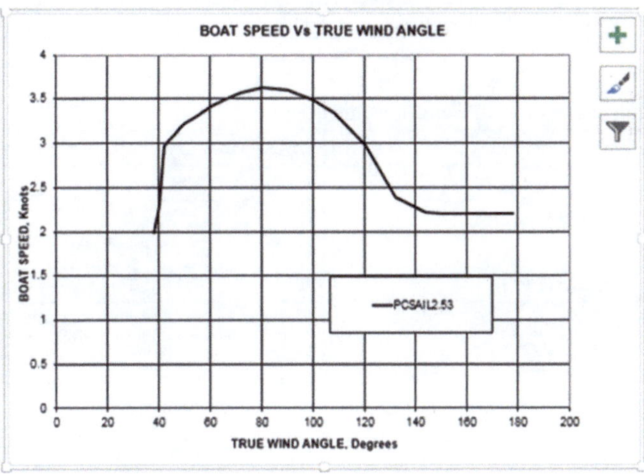

Fig. 2 PCSail boat speed estimate for MF in 6.6 knots of wind

forward, acquiring reliable data for the wind conditions and the corresponding boat speed was much more challenging. The VPP estimates are plotted as boat speed versus wind speed and true wind angle. The true wind angle is the angle of the wind relative to the boat when the boat is not moving, measured clockwise from the bow. For example, if the wind is coming from the starboard beam the true wind angle is 90°. This differs from apparent wind angle which includes the forward speed vector due to the vessel moving. In the beam wind example, if the true wind speed is 5

Table 1 Principal characteristics, *Mid-lenium FalcoN**

Dimension	Length (in)/Weight (lbs)
Length overall (LOA)	79
Beam (front. Keel)	12
Beam (behind, Keel)	12.5
Root cord (rudder)	4.25
Root cord (keel)	3.625
Freeboard – avg	4.75
Rudder span	25.5
Tip cord (rudder)	1.25
Bulb	15
Tip cord (keel)	1.625
Waterline length (LWL)	78.5
Draft	57.5
Draft max	57.5
Draft canoe	2
Water line (beam)	10
Displacement	52.7
Vert. Ctr. Of Grav. (VCG) from bulb bottom	−23.1
Long. Ctr. Of Grav. (LCG) from stem	44.6

Table 2 Sail dimensions for MF's medium air rig

Dimension	Length (in)
Main luff (P)	126
Boom Abv Deck (BAD)	11.38
Main foot (E)	46.75
Jib Luff Perpendicular (LPG)	26
(Jib hoist) I	114
Mast height (EHM)	133.5
Jib foot (J)	27.5

knots and the vessel is moving at five knots, the relative wind angle would be at 45°. Figure 1 shows an example plot for MF in 6.6 knots of wind. The plot shows her estimated top speed is 3.6 knots at a true wind angle of 80°. Note however, that a change from 120 to 130° true wind angle results in a boat speed drop from 3 to 2.4 knots, a 20% decrease, indicating the sensitivity to wind angle.

On-the-water testing was performed in the Severn River between College Creek and Spa Creek (Fig. 3) adjacent to the Naval Academy sea wall where the boat operators were standing. Relative wind angle readings were measured on board and were converted to true wind angles for the comparative analyses. Wind speed was

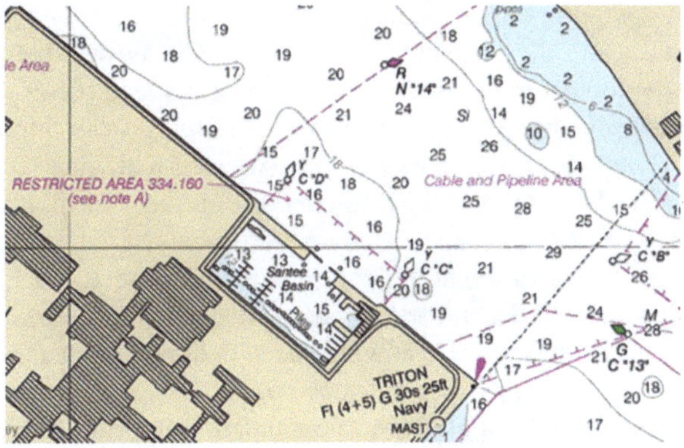

Fig. 3 Testing area. Predominantly near buoy D. The normal wind direction is southeast and north is at the top of the chartlet. The operators stood on the pier near the 15 foot sounding mark

measured by the operators on shore using a hand held anemometer at roughly at the height of the rig's center of effort. The boat generally sailed within 50 m of the operators. Some local building affects were likely when the wind was from the SSW to WNW. Luckily those wind directions were quite rare.

A specific point of sail was chosen for each data run. The boat was then sailed on a constant course and the sails were trimmed to avoid luffing. The time and wind speed was recorded and the course was held constant for at least 30 s. Data was collected at 1 Hz for time, position, relative wind direction, boat course, and boat speed using an onboard GPS tracking system linked to an onshore laptop placed near the testing area. Current was negligible (usually less than 0.1 knot, or 2.5% of typical boat speed). Following each run a different wind angle was chosen. Attempts were made with varying success to repeat each wind angle at least three times, however, setting the boat on a course to within a few degrees tolerance was challenging.

3 Results

Weather conditions were a significant barrier to acquiring data. To get good data the wind had to be between 5 and 15 knots and from a direction that was not affected by buildings or trees and the air temperature had to be in a safe working range for the personnel. During the winter testing was curtailed at dusk, which came as early as 4 PM. Testing began in September and ended in April, with about a third of the planned days resulting in good data. A total of 47 good data runs were completed.

Data reduction began with a review of the course and relative wind direction. Nearly constant values were needed to ensure the wind and course input to the VPP were accurate. Given the lightweight SailBot designs, small changes in wind speed or direction have a large impact on the boat speed. A run was considered for data reduction only if the course was constant within seven degrees and the wind speed within a knot.

After an experimental run was considered to yield consistent data the VPP was run using the measured wind speed and true wind angle. The predicted boat speed was then compared to the average boat speed recorded during the run. Reasonable data runs were completed in winds ranging from 5 to 10.6 knots.

Plots for boat speed were then created with the experimental data points collected from the data analysis and overlaid on the VPP plot. The experimental data points show the average speed and have error bars based on 95% confidence intervals. As the boat speed varied significantly with wind speed, the Coefficient of Variation (COV) was used in the uncertainty analysis to indicate variability. The minimum COV for boat speed was 1.4% at wind speed 6.2 knots with a true wind angle of 85°. The max COV was 35.5% at wind speed 6.0 knots with a true wind angle of 66°. The overall average COV of all experimental data runs was 19.4%.

Figures 4, 5, 6, 7, 8, 9, 10, 11, 12, 13, 14, 15, 16 and 17 are plots showing the VPP estimates with the experimental on-the-water data.

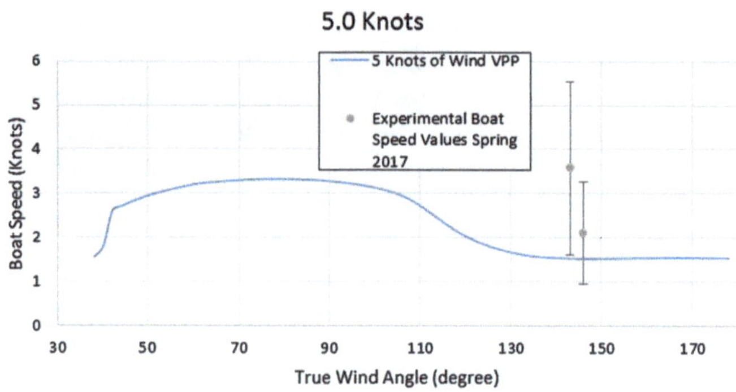

Fig. 4 Experimental result and VPP estimate at 5.0 knots wind speed

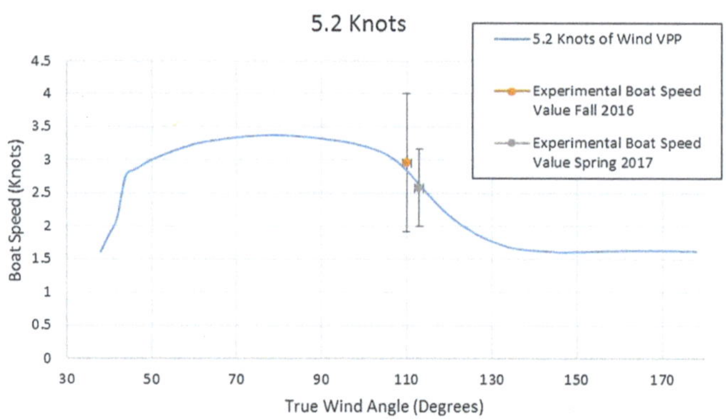

Fig. 5 Experimental result and VPP estimate at 5.2 knots wind speed

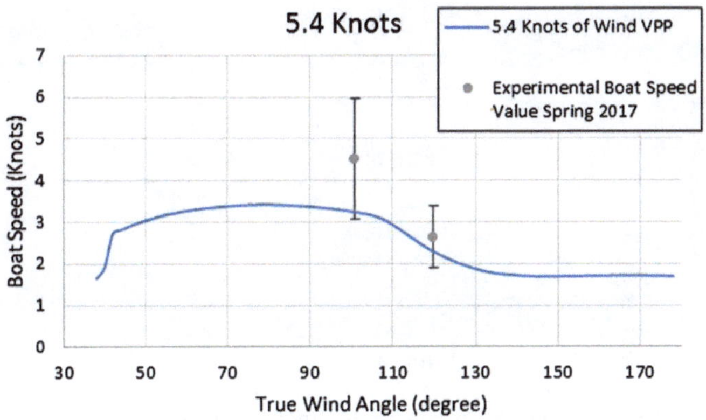

Fig. 6 Experimental result and VPP estimate at 5.4 knots wind speed

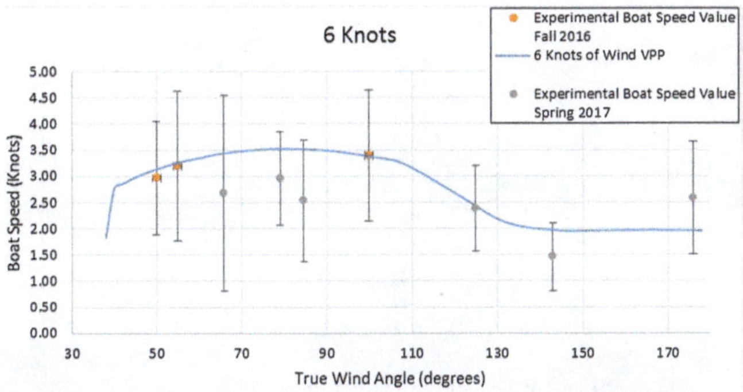

Fig. 7 Experimental result and VPP estimate at 6.0 knots wind speed

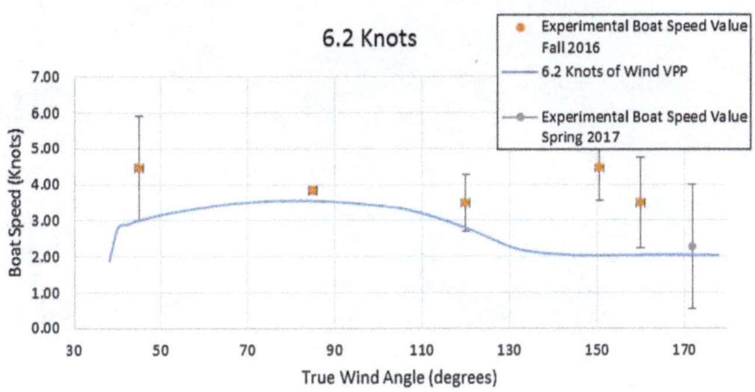

Fig. 8 Experimental result and VPP estimate at 6.2 knots wind speed

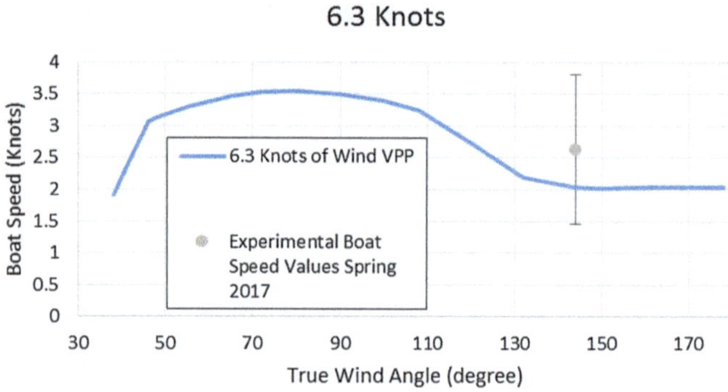

Fig. 9 Experimental result and VPP estimate at 6.3 knots wind speed

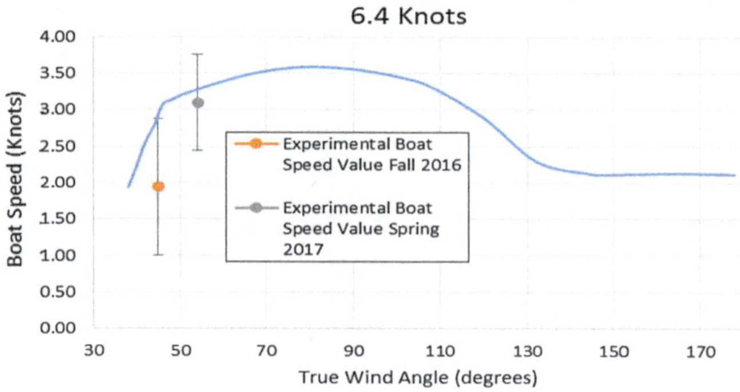

Fig. 10 Experimental result and VPP estimate at 6.4 knots wind speed

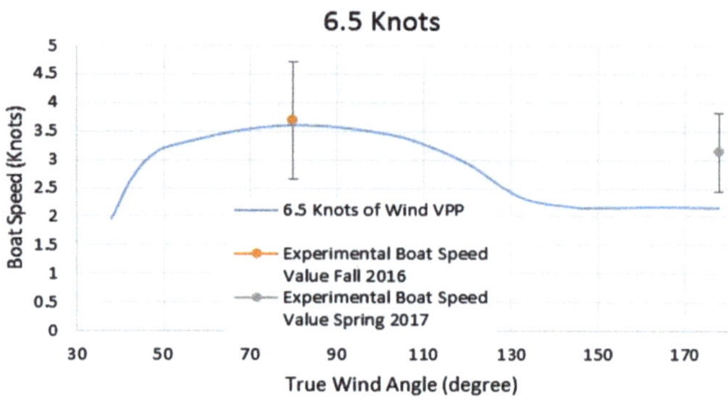

Fig. 11 Experimental result and VPP estimate at 6.5 knots wind speed

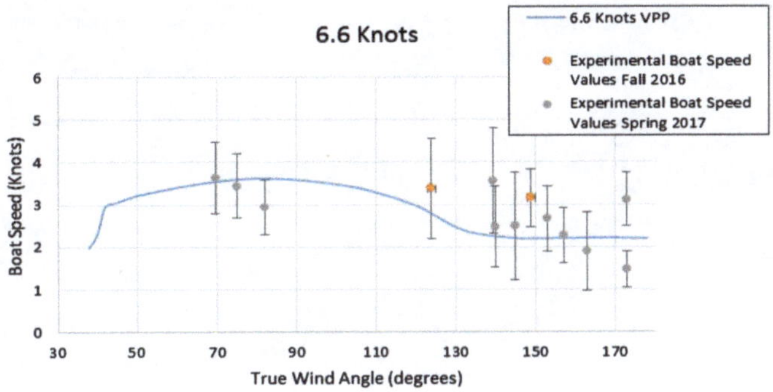

Fig. 12 Experimental result and VPP estimate at 6.6 knots wind speed

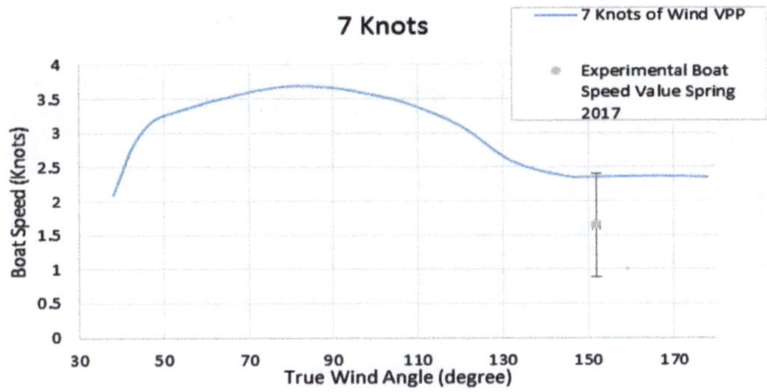

Fig. 13 Experimental result and VPP estimate at 7.0 knots wind speed

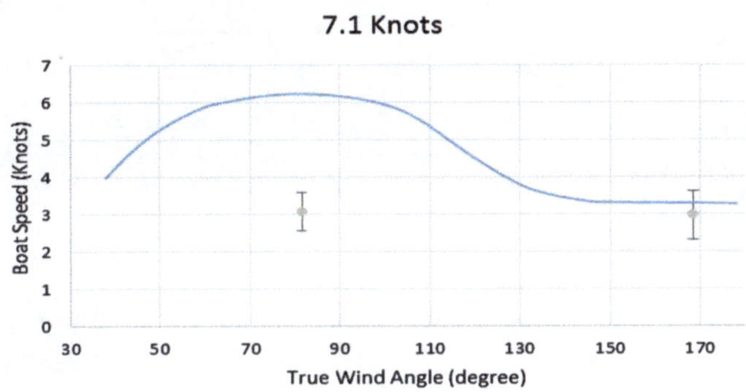

Fig. 14 Experimental result and VPP estimate at 7.1 knots wind speed

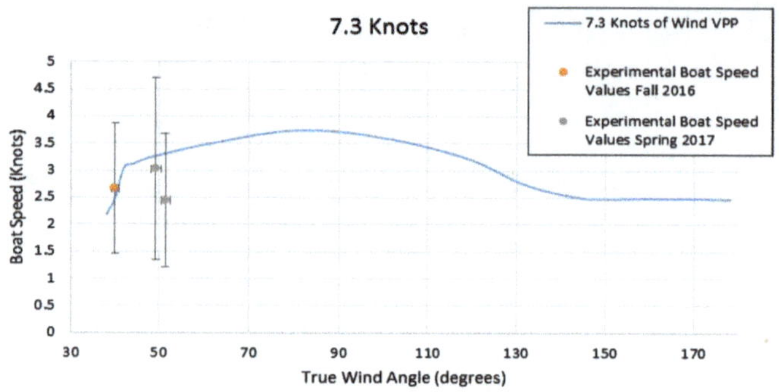

Fig. 15 Experimental result and VPP estimate at 7.3 knots wind speed

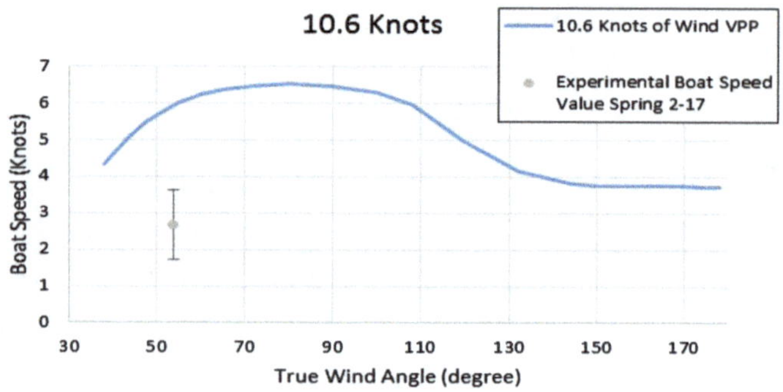

Fig. 16 Experimental result and VPP estimate at 10.6 knots wind speed

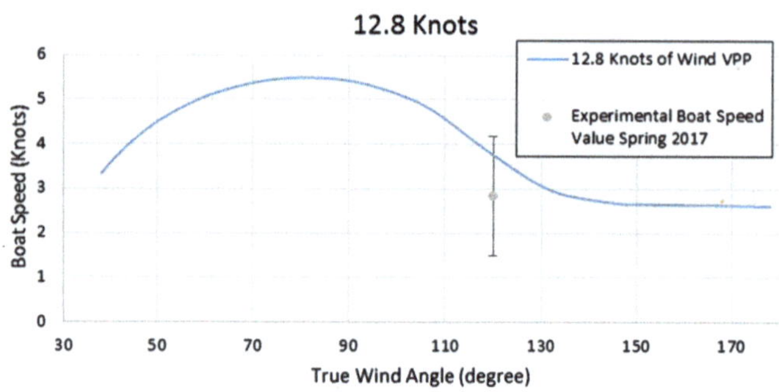

Fig. 17 Experimental result and VPP estimate at 12.8 knots wind speed

4 Discussion

Generally the VPP estimates fell within the 95% confidence level of the experimental data for wind speeds less than 7.3 knots with acceptable correlation below approximately 6.8 knots. Above that a trend then developed with the error bars gradually getting further from the VPP predations as the wind speed increased. Of the 45 data points below 7.5 knots, 25 had the average boat speed above the VPP estimate and 20 were below. Repeatability was generally good with the notable exception at a true wind angle of 173°. Even though the data was consistent within the two runs at that angle, the two resulting data points were far apart. The reason for this is unknown.

Much of the problem with the stronger winds was related to control problems. The medium air rig, while designed for winds up to 10 knots, was powerful enough to cause enough trim that the bow would submerge in winds as low as 8 knots. This significantly increased drag and caused directional stability issues that required large rudder angles to correct, therefore increasing drag significantly. Switching to the smaller sail area heavy air rig above 8 knots would improve control and probably the results.

Although the data was fairly consistent within a run, a number of variables likely influenced the scatter. Inconsistencies in the wind velocity and direction, even within a 30 s run, were observed and were likely caused by the proximity to land. Four different operators, with widely varying experience, drove and trimmed the sails, likely leading to inconsistencies in sail trim and rudder operation. The boat was stored unrigged, which meant that on the more than two-dozen days it was tested the rig was probably tuned slightly differently.

It would be nice to say that the on-the-water results completely validated the VPP. That however, is not the case from these results. While the data closely matched the predictions about half the time, the other half showed sometimes large differences. From these results it can only be said that the VPP seems to predict within the ballpark and has neither a slower or faster bias. The testing showed that getting good on-the-water data is challenging.

5 Future Recommendations

During the process of developing the test procedure and gathering the data, checklists were developed that significantly helped to maintain rigging consistency and that process should be continued with items such as sail trim. To avoid the land affect, the testing should be at least a few hundred meters from land, however that will require having a boat to carry the operators. Understanding the boat's limitations is important and future testing should understand the control limitations and limits of the different sail plans. For instance, the "medium air rig" should be replaced by the smaller "heavy air rig" in winds over 8–10 knots. Having a recording anemometer on the boat that is clear of the rig would provide more accurate wind speed data. The extra weight and windage can be factored in to the VPP. Although it was assumed that the operators could operate the boat more efficiently than the computer, it is likely the computer would be more consistent. This will require a greater refinement in our sail trim algorithm.

More statistical analysis is needed to determine if there are areas where the VPP might be improved. For example, an assumption going to the project was that the added resistance in waves component would under-predict the drag, leading to higher boat speed estimates than actual. The reasoning was that the component is based on freeboard versus wave height and the low freeboard SailBots see waves equal to their freeboard in winds as low as six knots. That issue did not seem to be the case however.

Acknowledgements We are very grateful for the help from Matt Hamlet who wrote the code we used to gather data and set up the control system. We also appreciate the timely response of CAPT David Jackson and Craig Patterson, who rescued MF when the battery died and she drifted away.

References

1. Gibbons-Neff, Peter, and Paul Miller. 2011. Route planning for a micro-transat voyage. In *4th IRSC*, Germany.
2. Larsson, L., R. Eliasson, and M. Orych. 2014. *Principles of yacht design*, 4th edn. Camden, Maine: McGraw Hill.
3. Martin, David E. and Robert F. Beck. 2001. PCSAIL, a velocity prediction program for a home computer. In *15th Chesapeake sailing yacht symposium*, 99–116.

APASail—An Agent-Based Platform for Autonomous Sailing Research and Competition

Bruno Alves, Bruno Veloso and Benedita Malheiro

Abstract This paper presents an Agent-based Platform for Autonomous Sailing—APASail—intended to support real and simulated competitions as well as assess navigation and boat modelling algorithms. The platform provides back-end services—competition server, boat modelling and data storage—and supports external browsers and software agents as front-end clients. The back-end adopts the Multi-Agent System (MAS) paradigm for the internal modelling of sailing boats and offers a Web Service Application Programming Interface (API) for the external software agents and a Web application for Web browsers. As a whole, the platform offers tracking (real competitions) and simulation (simulated competitions) modes. The testing and assessment of navigation algorithms and boat models correspond to private simulated competitions. In simulation mode, the back-end internal boat agent implements a simplified physical model, including the weight, sail area, angle of the sail and rudder, velocity and direction of the wind and position and velocity of the hull, whereas the front-end external boat agent implements the navigation algorithm on the team side, ensuring the privacy of strategic knowledge. The Web application allows the configuration and launching of competitions, the registration of teams and researchers, the uploading of boat physical features for simulation as well as the live or playback viewing of real and simulated competitions. The simulation mode is illustrated with the help of a case study. The APASail platform, which is open, scalable, modular and distributed, was designed for the research community to prepare, run and gather data from real and simulated autonomous sailing competitions.

B. Alves · B. Malheiro (✉)
School of Engineering, Polytechnics of Porto, ISEP-IPP,
Rua Dr. António Bernardino de Almeida, 431, Porto, Portugal
e-mail: mbm@isep.ipp.pt

B. Veloso · B. Malheiro
INESC TEC, Rua Dr. Roberto Frias, Porto, Portugal

K. I. Øvergård (ed.), *Robotic Sailing 2017*,
https://doi.org/10.1007/978-3-319-72739-4_3

1 Introduction

In recent years the research community has shown a significant interest in autonomous sailing competitions. The most important competitions involving autonomous sailing boats are the Microtransat Challenge [10], the World Robotic Sailing Championship (WRSC) [15] and the Sailbot competition [11].

The Microtransat Challenge [10] was created in 2005 as a transatlantic race for fully autonomous sailing boats comprising two classes: sailing boats propelled exclusively by wind power with hull load waterline length (LWL) restricted to 4 m; and non-sailing boats propelled by any other source of energy and with a LWL up to 2 m. One year later, the Sailbot International Robotic Sailing Regatta began at the Queen's University, Kingston, Ontario, Canada. This annual event takes place in North America and involves teams of students from different universities and colleges, including a 1 m class, a 2 m class and an open class for boats with a maximum length of 4 m [11]. In 2008, the World Robotic Sailing Championship (WRSC) and the International Robotic Sailing Conference (IRSC) started as a spin-off of the Microtransat Challenge. The WRSC contemplates the Micro-sailboat class, for boats with length overall (LOA) less than 1.5 m and a weight up to 100 kg, and the Sailboat class, for boats up to 4 m LOA and less than 300 kg [15].

Increasingly, researchers participate in these competitions in order to refine their navigation platforms and algorithms. However, since the logistics and costs of conducting real experiments with sailing boats are heavy and expensive, there is a lack of tools for the initial modelling, experimentation and debugging of sailing boat models and navigation algorithms. Moreover, such a platform can also be used to collect data for data mining as well as to follow in real time or playback competitions.

The goal of this project is to design, develop and test a versatile platform for simulated and real autonomous sailing competitions. In terms of general requirements, the platform should be open, scalable, modular and supported by open source technology. Additionally, it should implement the simulation and real operation modes. Users should be able to: (i) participate and follow live simulated competitions in simulation mode; (ii) follow live real sailing boat competitions in real tracking mode; and (iii) simulate their platforms and test their navigation algorithms in private simulation mode. Users include competition administrators and researchers (individuals or teams).

Our contribution is a platform providing a Representational State Transfer (REST) Web Service API for researchers to test navigation strategies and boat models, using private simulations, as well as to create, configure and follow in 2D real and simulated competitions, using a Web browser.

This paper is organized in five sections. Section 2 presents the related work. Section 3 details the project development. Section 4 presents a simulation mode case study. Finally, Sect. 5 concludes the paper, discussing the outcomes and future developments.

2 Related Work

This section provides an overview of open source sailing boat simulators, agent based competition platforms and autonomous sailing platforms.

Sailing Boat Simulators Sail7 is an open source tool designed for the analysis of sail boat performance. It was developed for XFLR5, an analysis tool for airfoils, wings and planes operating at low Reynolds Numbers, and is based on the Vortex Lattice Method and the 3D Panel Method [5]. Sail7 performs the polar analysis of the specified sail and hull combination.

Tracksail is an open source game developed in Java where players compete for the fastest time, guiding small sailing boats via the keyboard [8]. The implemented physical model contemplates the wind velocity and direction, the boat position, velocity and rudder angle.

Tracksail-AI, which is a modified version of the Tracksail sailing game, is intended to be played by Artificial Intelligence (AI) applications instead of humans [12]. It offers an open source API based on the original Java code to implement a simple sailing boat model, contemplating the wind velocity and direction, the boat position and velocity and the rudder and sail angles.

Multi-Agent Systems for Simulated Competitions RoboCup [6] is an annual international scientific competition in the areas of robotics and artificial intelligence. The RoboCup simulator consists of three large modules [7]: client—responsible for the dynamics of its team players and communicates with the server module via a socket, sending messages (messages) controlling the actions of the players through the protocol of the transport layer User Datagram Protocol (UDP); server and visualization.

The Simulated Car Racing Championship (SCRC) [9] is a competition platform based on the The Open Racing Car Simulator (TORCS). TORCS [3, 9, 16] provides a way to create, test and use simulated cars (BOT). Each running BOT consists of a server and a corresponding client, communicating via a dedicated application layer protocol over UDP. The BOT server sends the status of the sensors (perceptions obtained from the simulated environment) to the client every 20 ms and the client has 12 ms to decide and communicate the BOT server which actions to take. The client implements locally and independently the control algorithm.

The Trading Agent Competition (TAC) [14] involves eight trading agents in an auction-based marketplace. The main purpose of the agents (travel agents) is to create travel packages—a round-trip flight, a five-day stay in a hotel and entertainment—according to the preferences of their clients. Each agent communicates with the TAC-Server through messages on Transmission Control Protocol (TCP). The server provides a specific API for clients to access the market information and submit proposals.

Autonomous Sailing Platforms Several researchers have designed computational platforms for autonomous sailing addressing the problems of tracking and collision avoidance [2], testing of sailing missions [1] or AI boat control algorithms [4] as well as real and simulated boat control and modelling [13].

In 2011, Ammann et al. proposed a world server approach for the WRSC competition in order to gather, broadcast and store data from all vessels and, thus, provide tracking and collision avoidance functionalities [2].

Davison, in 2013, developed YachtSim, a yacht simulator written in C#, to aid in developing AI algorithms for an autonomous yacht [4]. The code was inspired by TrackSail-AI and includes three main applications: YachtSimServer, YachtSimClient and EyeInTheSky. The YachtSimServer provides the interface between the virtual yacht hardware and the AI running on the YachtSimClient and includes a list of regatta environment simulators. The EyeInTheSky is a client module responsible for the visual display of where and how the virtual yacht is sailing.

In 2015, Alves et al. proposed a simulation platform for testing sailing missions called MetaSail [1]. It takes into account the boat's heeling effect on the sail forces and the leeway angle, along with the righting moment produced by the keel and the hull's hydrostatic forces.

Taylor et al. designed in 2016 the experimental robotic sailing boat daemon (boatd) [13] to support real and simulated boat control, modelling, telemetry and logging. Boats are externally controlled via an HTTP API and the commands are forwarded through drivers to the Tracksail-AI simulator or to the controller of an autonomous sailing boat.

Our Agent based Platform for Autonomous Sailing Research and Competition (APASail) provides a dual mode—real and simulated—competition platform for autonomous sailing boats, offering a REST Web Service API and a Web application. The 2D simulation mode is based on Tracksail [8] and, additionally, implements the influence of the mass and the sail angle & area. Furthermore, it supports the test and simulation of navigation algorithms and boat models as private simulations.

3 APASail

Use Cases The contemplated use cases were: (i) registration and authentication; (ii) creation of real and simulated public competitions; (iii) creation of private simulations to test navigation algorithms; (iv) competition tracking; and (v) platform administration.

System Architecture Fig. 1 displays the overall system architecture comprised of front-end and back-end. The back-end includes, in terms of components, two database servers (PostgresSQL and RethinkDB), a Web server (Apache), a JavaScript Object Notation (JSON) server (Node.js), an agent execution platform (JADE) and a RESTful API framework (RESTlet). The Apache HyperText Transfer Protocol (HTTP) server deploys the Web application, which was developed using Boostrap. The Node.js, while a server-side JavaScript module, automatically refreshes the client Web pages and acts as a bridge between the client browser and both databases. Finally, the Java Agent DEvelopment (JADE) and the Restlet frameworks are the core of the platform, i.e., provide support to simulated and real competitions. Each

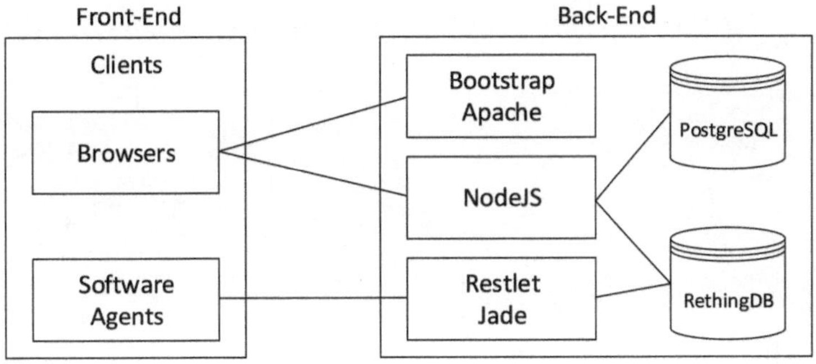

Fig. 1 System Architecture

competition is modelled as a multi-agent system, corresponding to a JADE container, where the participating boats (simulated or real) are modelled as JADE agents. The interaction between the JADE platform and the competitor boat agents is achieved through the implemented REST interface. The JADE platform is responsible for communicating all live competition data to the RethinkDB database.

The front-end contemplates two types of clients: browsers and software agents. The browser obtains the Web pages from the Apache HTTP server and interacts with the Node.js using WebSockets to update and control Web page contents. Competitors use browsers for user registration and authentication, team and boat profiling, tracking real time competitions (real and simulated) as well as to define and follow tests. In addition, administrators can manage users, teams and competitions. The client agents execute on the competitor platform. Each client agent represents an autonomous sailing platform and interacts with its JADE representative using the implemented JADE Restlet API. While in simulation mode, the simulated boat agent runs the navigation algorithm, in real mode the boat agent executes the navigation algorithm on board of the platform. In both modes, the boat agent communicates with its JADE representative to report its status and location and receive the competition status.

Functionalities In terms of permanent data modelling and storage, there are two databases. PostgreSQL database stores all the information related with users, teams, boats and competitions. The RethinkDB database stores real time competition data, i.e., all competition events, including weather conditions and the position, velocity and time (PVT) of the boats.

The Web application allows users to register, authenticate and manage the profile, create private competitions and participate in real and simulated competitions. Depending on the user role, the user will be able to create (administrator), configure boat and team profiles (administrator and competitor), create, configure, follow live or playback real and simulated competitions (administrator and competitor), using an interactive 2D graphical interface.

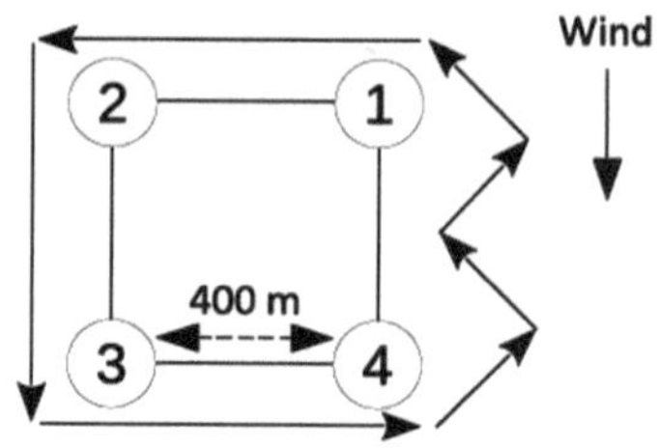

Fig. 2 Course layout used in the simulation tests

4 Simulation Case Study

In order to test the adopted physical modelling approach, a simulation mode case study, involving two sailboats ($boat_A$ and $boat_B$), was performed. Figure 2 displays the sail course layout defined by four marks (buoy 1 to buoy 4) with a length of 400 m/leg. The tests contemplated: (i) beating between the buoy 4 and buoy 1; (ii) beam reach between buoys 1–2 and 3–4; (iii) running between buoys 2–3; and (iv) the influence of the mass and sail area. The true wind velocity and direction were set to 9 kn $\pm$1 kn and 90°$\pm$10°, respectively. The boats have a mass of 200 kg, a length of 3.7 m, a LWL of 3.2 m and a sail area of 3.4 m^2 in the case of $boat_A$ and of 3.0 m^2 in the case of $boat_B$.

Beating In this test, which corresponds to the first leg of the course, $boat_A$ was set to sail at an angle of 45° to the wind, while $boat_B$ adopted an angle of 40° to the wind. $boat_A$ took 450 s at an average velocity of 2.33 kn, whereas $boat_B$ spent 729 s at an average speed of 1.58 kn. As expected, $boat_A$ is faster than the sail boat $boat_B$.

Beam Reach This test, which corresponds to the second and forth legs of the course, was performed with both boats navigating perpendicularly to the wind. The average time and velocity in these legs was 234 s and 3.24 kn for $boat_A$, and 249 s and 3.03 kn for $boat_B$. Not only the boats sailed faster than in the first leg, but $boat_A$, with larger sail area, was the fastest.

Running In this test, which corresponds to the third leg of the course (buoy 2–buoy 3), both boats are running downwind. The average time and velocity in these legs was 180 s and 4.18 kn for $boat_A$ and 189 s and 3.98 kn for $boat_B$. As expected, this was the fastest leg and $boat_A$, with larger sail area, was again the fastest.

Boat Mass and Sail Area The influence of the mass and the sail area on the velocity of a sail boat was tested using the boat polar diagrams generated from the implemented boat physical model. In this case, the wind velocity remained at 10 kn, while the boat direction varied between 0° and 360°. Figure 3a represents the velocity polar diagrams of two sail boats with equal LWL (3.2 m) and sail area (3.4m^2), but different mass ($boat_A$ with 140 kg and $boat_B$ with 200 kg). As expected, the lightest sail boat is faster than the heavier sail boat. Figure 3b shows the velocity polar diagram of two

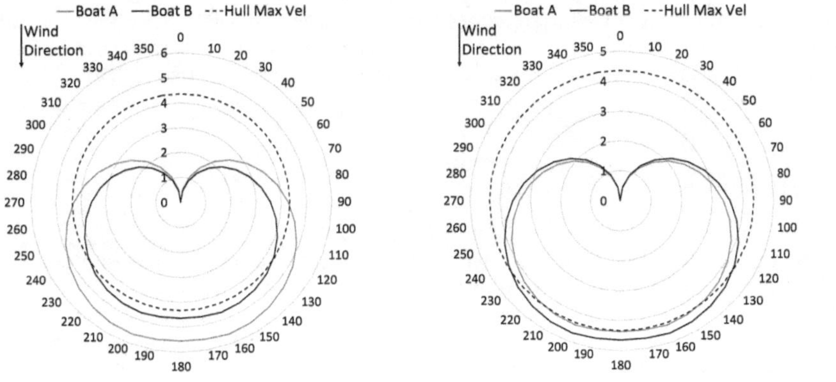

(a) Different weights. (b) Different sail areas.

Fig. 3 Polar Diagrams

sail boats with equal equal LWL (3.2 m) and mass (200 kg), but different sail area ($boat_A$ with 3.0 m^2 and $boat_B$ with 3.4 m^2). As expected, the boat with larger sail area is faster. The polar diagrams display also the maximum hull velocity based on the LWL of the boats.

5 Conclusion and Future Work

APASail supports real and simulated autonomous sailing competitions and interfaces seamlessly with external sailing boat controllers, client navigation agents and browsers. Internally, this research and competition platform uses dedicated agents to model sailing boats, i.e., to process real time events in real and simulation modes and implement the physical modelling in simulation mode. The latter takes into account the position, velocity, mass, rudder angle, sail area and sail angle of the sailing boat as well as the direction and velocity of the wind. When compared with Tracksail, our model additionally contemplates the influence of the mass and sail area & angle. Furthermore, researchers can create, participate and follow (live or playback) real and simulated public competitions in 2D as well as create private simulations to test different boat models (implemented in JADE agents) and navigation strategies (implemented in client agents).

In the future, we intend to improve the physical modelling in order to contemplate the boat LWL, wave height, tide and current as well as test the platform thoroughly both with real and simulated competitions. Finally, we would like to explore with 3D boat rendering.

Acknowledgements This work was partially financed by the European Regional Development Fund (ERDF) through the Operational Programme for Competitiveness and Internationalisation

(COMPETE Programme), within project «FCOMP-01-0202-FEDER-023151» and project «POCI-01-0145-FEDER-006961», and by national funds through the Fundação para a Ciência e Tecnologia (FCT)—Portuguese Foundation for Science and Technology—as part of project UID/EEA/50014/2013.

References

1. Alves, J.C., and N.A. Cruz. 2015. METASail—A tool for planning, supervision and analysis of robotic sailboat missions, 57–64. Cham: Springer International Publishing.
2. Ammann, N., F. Hartmann, P. Jauer, J. Krüger, T. Meyer, R. Bruder, and A. Schlaefer. 2011. Global data storage for collision avoidance in robotic sailboat racing—the World Server Approach, 157–166. Berlin: Springer.
3. Corneliussen, S.C., and M. Westergaard. 2011. Combining offline and online learning in developing an adaptive controller for a simulated car racing environment.
4. Davison, B. 2016. YachtSym. https://github.com/rbdavison/YachtSim. Accessed June 2016.
5. Deperrois, A. 2016. Sail7. http://www.xflr5.com/sail7/sail7.html. (2012). Accessed June 2016.
6. Federation, T.R. 2015. About robocup: What is robocup?. http://www.robocup.org/about-robocup/.
7. Kitano, H., M. Asada, Y. Kuniyoshi, I. Noda, E. Osawa, and H. Matsubara. 1997. Robocup: A challenge problem for AI. *AI Magazine* 18 (1): 73–85.
8. Kuusela, T., S. Brockmann, R. Johnson. 2005. Tracksail. http://tracksail.sourceforge.net/. Accessed June 2016.
9. Loiacono, D., L. Cardamone, P.L. Lanzi. 2013. Simulated car racing championship: Competition software manual. http://arxiv.org/pdf/1304.1672.pdf. Accessed June 2016.
10. Microtransat.org. 2016. The microtransat challenge. http://www.microtransat.org/. Accessed June 2016.
11. Sailbot.org. 2016. Sailbot. http://sailbot.org/. Accessed June 2016.
12. Taylor, L., S. Nicholls, L. Stone, T. Cartwright. 2013. TrackSail-AI. http://github.com/boatd/tracksail-ai. Accessed June 2016.
13. Taylor, L., S. Nicholls, L. Stone, T. Cartwright. 2016. Experimental robotic sailing boat daemon. http://boatd.readthedocs.io/. Accessed April 2017.
14. Wellman, P., P. Wurman, K. O'Malley, R. Bangera, S.D. Lin, D. Reeves, and W. Walsh. 2001. Designing the market game for a trading agent competition. *IEEE Internet Computing* 5(2): 43–51.
15. WRSC Organising Committee. 2016. World robotic sailing championship. http://www.wrsc2016.com/competition.html. Accessed June 2016.
16. Wymann, B., C Dimitrakakis, A. Sumner, E. Espi, C. Guionneau, R. Coulom. 2013. TORCS, the open racing car simulator. http://www.torcs.org. Accessed June 2016.

A Component-Based C++ Communication Middleware for an Autonomous Robotic Sailboat

Francisco J. Santana-Jorge, Antonio C. Domínguez-Brito and Jorge Cabrera-Gámez

Abstract The new C++ standard, C++11, and its upgrade, C++14, introduces new advances and features which make more affordable and easier the development of software for complex systems. Following this tenet we have designed and developed a component-based service-oriented C++ middleware, called ISE, for distributed systems using exclusively standard C++ and the quasi standard C++ Boost Libraries for keeping the middleware portable. The final aim of developing ISE has been to build the remote communication software infrastructure of an oceanic autonomous robotic sailboat called A-Tirma.

1 Introduction

The development of an autonomous sailboat is a demanding endeavor that is not circumscribed to the vessel. Normally, it implies also the development of a communication infrastructure which may consume a considerable amount of resources. The complexity of this communication infrastructure is commonly overlooked but it is undoubtedly a central piece of the whole system, which may become quite complex in terms of software engineering. In this paper we introduce a component-based C++ middleware called ISE (Integrating Software Entities), we have developed specifically for building the communication software infrastructure for an autonomous sailboat. We have used a CBSE (Component-Based Software Engineering) paradigm [1] and a service-oriented architecture [2] to define a service-oriented component model. In ISE, a robotic system is a distributed system [3] built up of components, where each component provides and/or requires services to/from others. Moreover, it is possible to dynamically reconfigure a whole system, as ISE provides support for

F. J. Santana-Jorge · A. C. Domínguez-Brito (✉) · J. Cabrera-Gámez
Instituto Universitario SIANI, Universidad de Las Palmas de Gran Canaria, Las Palmas, Spain
e-mail: antonio.dominguez@ulpgc.es

A. C. Domínguez-Brito · J. Cabrera-Gámez
Departamento de Informática y Sistemas, Universidad de Las Palmas de Gran Canaria, Las Palmas, Spain

K. I. Øvergård (ed.), *Robotic Sailing 2017*,
https://doi.org/10.1007/978-3-319-72739-4_4

on-demand dynamic component instantiation, and dynamic runtime type information about a system and its components. As case application in a real scenario, we have implemented using ISE a first operative version of a distributed software infrastructure for remote control and monitoring of the autonomous sail vessel A-Tirma [4] developed at our laboratory. In the rest of the document we describe in more detail the middleware, how we have applied it so far to our projet A-Tirma, and the conclusions drawn from the work underdone.

2 ISE Middleware

As main design principle for developing ISE we highlight the use of *standard and portable C++* as implementation language. C++ is a systems programming language that offers good use of hardware, effective abstraction and allow for real time performance [5]. With the appearance of new C++ standards (C++11 and C++14, implemented by most mainstream compilers [6]), the language has been modernized with many new features (lambda functions, move semantics, etc. [7]) which makes C++ code simpler. In addition to guarantee portability, we have also used Boost C++ libraries [8] given their "quasi-standard" status within the C++ community. Since software components, as composition units to integrate systems, provide a higher level of encapsulation, abstraction decoupling and reuse for software systems, a *component based* middleware [1, 9, 10] has been designed and developed. Moreover, the middleware is *service-oriented*, that is, component's external interfaces will be defined as services provided and/or required [2].

In Fig. 1 we can appreciate the distributed layered architecture of ISE, organized in two layers. The first layer, *the connection manager*, is responsible for managing peer-

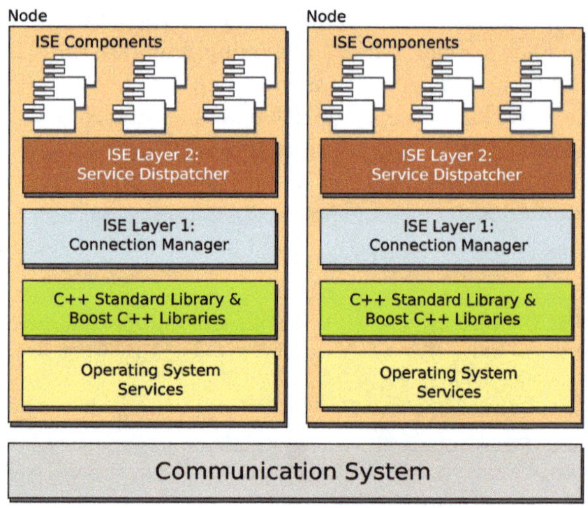

Fig. 1 ISE layered architecture

to-peer connections amongst components, which isolates users and developers from the low level details for handling asynchronous connections using both TCP and UDP protocols. The second layer, *the service dispatcher*, abstracts the dispatching and housekeeping of all services provided by ISE components in order to carry out their specific functionalities. Those two layers support the run-time software infrastructure and the application programming interface (API) the middleware provides to ISE software components, as we can see in the figure.

Software components (components for short) are distributed *active objects* [11], in the sense that each of them has its own flow of execution, mapped as operating system processes or threads in a distributed system. In ISE, the external interface of a component is the enumeration of the services it provides (*provided* services), and also of the *potential* services it might use or require from others (*required* services). Therefore, any middleware interaction amongst components is abstracted by using services, and, to keep things simple, there are only two generic types of services. Namely, *request-response* and *subscription* services.

Request-response services. The sequence diagram of Fig. 2 illustrates how a request-response service operates. A service is started with a request addressed to the provider by the requester. The provider answers with a response according to the request received. Although not shown in the figure for clarification reasons, in ISE, on the requester side, the component might do a blocking or an asynchronous wait for the response. In the provider side service requests are received and processed asynchronously, usually associated to *callbacks* which are user-defined. This type of service put into practice a *pull* model of communication.

Subscription services. Figure 3 depicts the sequence diagram of a typical subscription service in ISE [10]. It implements a publish/subscribe paradigm of communication. A service is initiated by the requester by sending a subscription request. Once subscribed, the provider will start sending *subscriptions* (published data) to the requester, either periodically or in an asynchronous event basis. In normal conditions, the provider will keep serving subscriptions until the requester explicitly requests to get unsubscribed. Several callbacks are associated to both sides. On the requester side, to subscriptions when received. On the provider side, to requests for getting subscribed or unsubscribed, and to publish data periodically using timers or events. Subscription services provide a *push* model of communication.

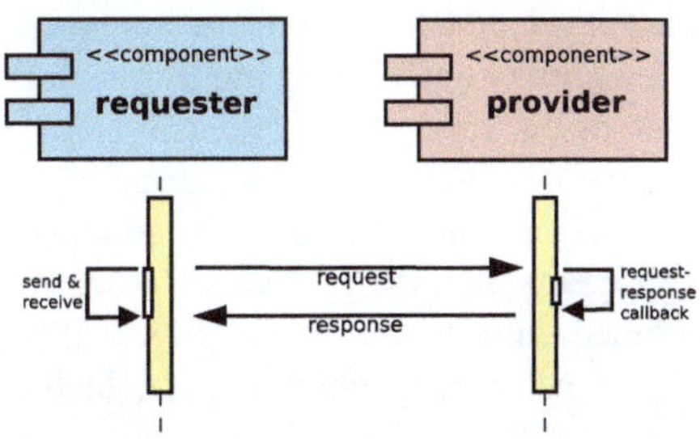

Fig. 2 Request-response service

Fig. 3 Subscription service

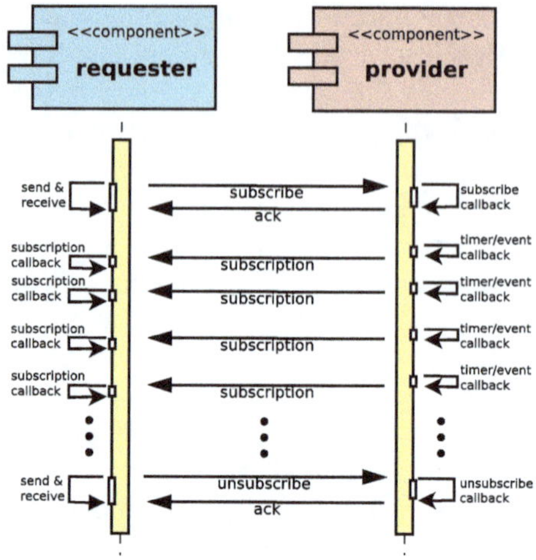

To implement both types of generic services in ISE we have made use of Boost C++ library Boost.Asio [12], a library for network and low level I/O programming. In particular, ISE services component interactions are carried out sending packets through TCP connections or UDP datagrams between the components involved.

More specifically, in ISE, service interactions amongst components are done transferring discrete units of information we name *packets*. Furthermore, the services a component provides are defined by the type of packets each service uses. To provide generality and complete freedom of implementation to component developers, component's services and service type packets are all user-defined in ISE. Thus, for response-request services two types of packets define a service: a *request* packet and a *response* packet, shown in Fig. 2. Likewise, subscription services are defined by five types of packets, namely: two packets for getting subscribed (request and response), a packet for *subscriptions*, and two packets for getting unsubscribed (request and response), all of them shown in Fig. 3.

ISE packets are easily defined directly in C++ code as Fig. 4 illustrates. Each type of packet is mapped as a C++ class. Packet data members can be defined to have any of the following types: C++ primitive types, `std::string`, any other ISE packet type and the container type `ise::packet::containers::vector`. As to packet's marshaling and unmarshaling the serialization interface for packets has been implemented using the Boost.Serialization Library [13]. Figure 4 shows examples of packets used in a real system. Concretely, packets `subscription_rq`, `subscription_rs`, `unsubscription_rq` and `unsubscription_rs` are definitions shared by several distinct subscription services. Services are defined in a similar manner. In Fig. 5 we show an example of a subscription service, `xb_local`, defined with the packets of Fig. 4. ISE packets and services are usually defined in C++ header files, which may be included wherever necessary. The complete code behind ISE packets and services is generated using macros like the ones appearing in the previous figures, which have been implemented using the Boost.Preprocessor

Fig. 4 C++ Packet definitions for a subscription service

```
ISE_DEFINE_PACKET_IN_NAMESPACE(
  ADD_NAMESPACE( a_tirma_tools ),
  subscription_rq,
  ADD_MEMBER( bool, status )
)
ISE_DEFINE_PACKET(
  subscription_rs,
  ADD_MEMBER( bool, status )
  ADD_MEMBER( std::string, ticket )
)
ISE_DEFINE_PACKET_IN_NAMESPACE(
  ADD_NAMESPACE( a_tirma_tools ),
  unsubscription_rq,
  ADD_MEMBER( std::string, ticket )
)
ISE_DEFINE_PACKET_IN_NAMESPACE(
  ADD_NAMESPACE( a_tirma_tools ),
  unsubscription_rs,
  ADD_MEMBER( bool, status )
)
ISE_DEFINE_PACKET_IN_NAMESPACE(
  ADD_NAMESPACE( a_tirma_tools ),
  xb_local_sb,
  ADD_MEMBER( int, power)
  ADD_MEMBER( int, temperature)
  ADD_MEMBER( int, rssi)
  ADD_MEMBER( int, failing_deliveries)
)
```

Fig. 5 C++ service definition for a subscription service

```
namespace a_tirma_tools
{
  ...

  ISE_PROVIDED_SERVICES(
    a_tirma_xb_services_to_provide,

    ... // rest of services

    SUBSCRIPTION_PROVIDED_SERVICE(
      subscription_rq,
      subscription_rs,
      unsubscription_rq,
      unsubscription_rs,
      xb_local_sb,
      xb_local // service
    )
  )
}
```

Library [14]. The use of this library has allowed to utilize C++ itself as IDL (Interface Definition Language) for defining interfaces between ISE components, in contraposition to other middlewares where usually IDL code must be compiled to generate its corresponding code in the final implementation language, (C, C++, Java, Python, etc.). We consider this an interesting simplification that facilitates middleware utilization and reduces its learning curve.

ISE provides a C++ class which embodies what is a component in the middleware, class `ise::component`. Using this class, either through derivation or through composition, we can define our own components endowed with its particular functionalities. As illustrated in Figs. 2 and 3, in a given component, the functionality provided by a specific service is defined using *callbacks*, namely: function objects, lambdas, function pointers, etc. And it is through services that we define the behavior and functionality of an ISE component. Thence, for request-response services, in the provider side we associate a callback for the request packet. And similarly, for subscription services, in the provider side, we associate callbacks to requests for getting subscribed or unsubscribed. Accordingly, in the requester side, we associate a callback for the subscription packets it will receive when subscribed.

As general services, the middleware provides services for remote instantiation and name resolution. More concretely, it is possible to instantiate remotely ISE components using the services offered by a type of component provided by the middleware itself (component *instantiation_server*). We have used Boost C++ library Boost.Process [15] to implement ISE's instantiation services. In addition, the middleware also includes a type of component, called *name server*, which provides name resolution services. In fact, name servers allow to register information related to how to locate component instances, the services (and its packets) they provide, and, if any, about the instantiation servers present in a system as well. They allow also to define component namespaces, in order to organize logically a distributed system. Furthermore, name servers are also part of the infrastructure which allows to instantiate packets dynamically. Indeed, the Boost C++ library Boost.Any [16] has been used to implement the dynamic type safe behavior of ISE packets.

3 A-Tirma's Communication Infrastructure

We have designed and developed ISE for supporting the software communication infrastructure of an oceanic autonomous sailboat [4], developed at our laboratory, called *A-Tirma* (shown in Figs. 6 and 7). Figure 8 displays the deployment diagram of the embedded system on-board A-Tirma. The hardware of the main controller system is an Atmel ATSAM3X8E microcontroller based Arduino DUE prototyping platform. This main controller is in charge of sailboat autonomous navigation and short range communications via an XBee 868 Pro radio link [17]. The autonomous navigation control software it executes runs on ChibiOS [18] (more details in [19]). A second controller based on a Waspmote [20], an Atmel ATmega1281 microcontroller based prototyping board, is responsible for long-range communications and last resource localization signaling in case of emergency. The communication controller acts as a gateway for sending and receiving, telemetry and remote telecommands. It uses a 3G/GPRS module[1] for data communications via the mobile phone network, and a Rockblock 9602 transceiver[2] for satellite communications utilizing the Iridium SBD (Short Burst Data) service.[3] XBee radio communications are used at short range (about 1–2 kms.) and real-time communication with the autonomous vessel. GPRS mobile communications have been included to communicate with the vessel at long range when situated in areas of mobile network communication coverage, typically coastal areas. Finally, for oceanic long range communication where mobile communications are not possible, the communication with the sailboat is based on Iridium SBD satellite communications. In addition to those three types of

[1]Waspmote GPRS + GPS Module (https://www.cooking-hacks.com/gprs-gps-module).

[2]RockBLOCK Iridium SDB 9602 Transceiver (http://www.rock7mobile.com/products-rockblock).

[3]Iridium SBD (https://www.iridium.com/services/details/iridium-sbd).

Fig. 6 A-Tirma G2

Fig. 7 A-Tirma in the water

communications links, we can command directly the rudders and the sails of our vessel using a remote-controlled device (through an RC receiver). This link is wired at low level for direct control of the vessel's actuators and usually it is used when the sailboat is at sight. Mind, as well, that using GPRS and Iridium SBD links implies paying for the communications, whether the mobile telephone network fee, or the Iridium satellite access and data fees, being significantly more expensive the latter ones.

Figure 9 depicts the deployment diagram for the communication infrastructure we have developed using ISE. Mind that the middleware has been not used on the sailboat, as the microcontroller platforms on-board are not computationally powerful enough, specially for their lack of memory. Indeed, we have selected them mainly for their low power consumption requirements ([4, 21]). In fact, neither the

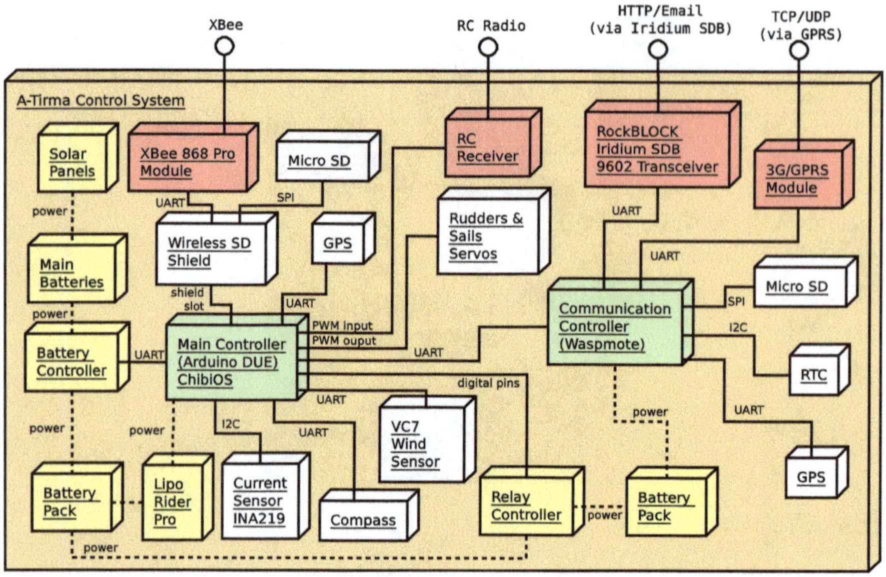

Fig. 8 A-Tirma's embedded control system. Deployment diagram. Power connections between elements are shown as discontinued lines, for clarification reasons not all of them are shown. For the same reasons, physical switches are also not shown in the diagram

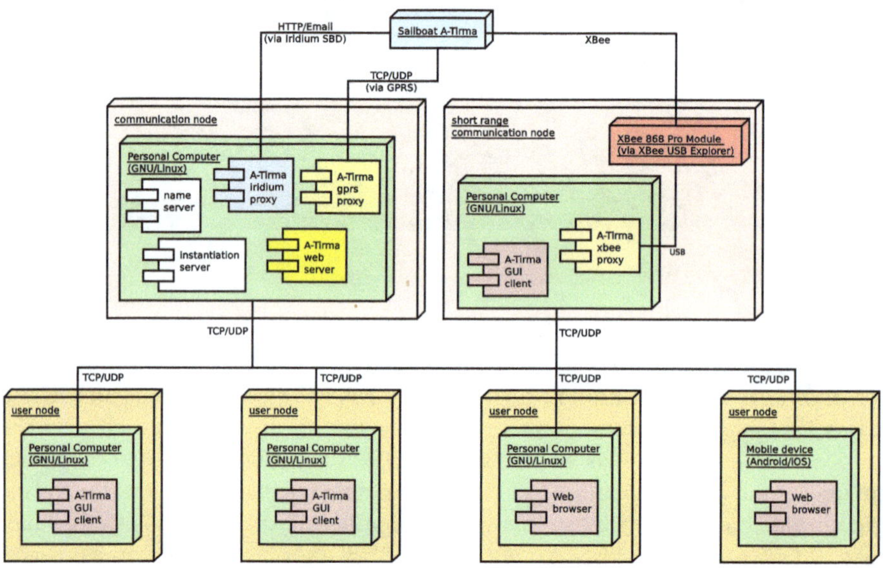

Fig. 9 A-Tirma's communication infrastructure. Deployment diagram. To keep the diagram as clear as possible, connections between ISE components are not shown. All ISE services are provided via TCP/UDP, so, implicitly, any component can access any service provided by any component

main controller nor the communication controller have support for the C++ Standard Library, or the C++ Boost Libraries, which are necessary for running the middleware. The figure shows a typical system deployment in a navigation mission. As we can observe, there are four types of nodes: the autonomous sailboat itself, a *short range communication* node, a *communication* node for long range communications and several *user* nodes. Tables 1 and 2 describe briefly the ISE services and components building up the system. To keep Fig. 9 simple, ISE connections between component are not shown. Note that all ISE services are provided through TCP or UDP protocols (this is specified at component instantiation), components implicitly connect to their service providers as specified in Tables 1 and 2.

As we can observe in Fig. 9, usually a *short range communication* node is deployed in a typical mission scenario, specially if we will be supervising and monitoring the navigation of the boat at short distance using the XBee radio link. The hardware we usually utilize for deploying this node is a notebook under GNU/Linux Ubuntu, or a Raspberry PI under Raspbian exporting its graphical interface via VNC on a 10"-Android-based tablet. This node normally host a XBee proxy component and a GUI client component. The *communication* node typically is deployed on a personal computer under GNU/Linux Ubuntu in our lab, and usually host the proxy components we need for long range communications, namely the GPRS proxy component and the Iridium SBD proxy component. In addition, it ordinarily hosts also a web server component which uses the services of all proxy communication components (XBee, GPRS and Iridium SBD ones) for publishing the sailboat's tracking telemetry information through a web page. As to the *user* nodes appearing in Fig. 9, two of them host mainly a GUI client component instance. Those components use the services provided by the proxy communication components running in the communication nodes and implement a remote interface for monitoring and control of the sailboat. A snapshot of the interface provided by GUI client component instances is shown in Fig. 10. They may be instantiated at different user nodes, so several users may be remotely controlling and monitoring the vessel at any given moment. Using them we can monitor and control many aspects of the embedded system onboard vessel A-Tirma. Thus, for example, it is possible to modify parameters affecting the sailboat behavior during autonomous navigation like time to next tacking, forbidden angle range to navigate up and down wind, period of telemetry messages, whether activating leeway compensation or not, adding/removing navigation waypoints, changing vessel's navigation mode, etc. just to name a few. The hardware we have ordinarily use for running user nodes is a typical personal computer under GNU/Linux Ubuntu. Furthermore, using the web service provided by the web server component hosted in the communication node, we can access a tracking web page during a mission, using a typical personal computer, or a mobile device like a smartphone or tablet (Fig. 11 shows a snapshop of this tracking web page).

In order to test the communication infrastructure, amongst the different real on-field tests we have carried out, in Fig. 12, we provide the tracking log of a 10-h stress test which took place in the bay of El Confital in Gran Canaria. In this stress test A-Tirma kept navigating autonomously during about 9–10 h of continuous navigation following consecutively two sailing triangles. The objective of the test was to verify

Table 1 A-Tirma communication system. ISE services

Services			
Service	Type	Provided by	Description
`telemetry`	subscription	XBee, GPRS and Iridium SBD proxy components	Provides periodically telemetry information to subscriber components. Subscription telemetry packets are sent to subscribers in raw format, as they come from the boat, subscribers are responsible for parsing telemetry packets. There are several telemetry types depending on the information they transport (location, waypoints, configuration parameters, etc.)
`telecommand`	request-response	XBee, GPRS and Iridium SBD proxy components	Service provided to send telecommands to the sailboat. Telecommand packets are also sent in raw format (the one accepted by the sailboat). There are also distinct types of telecommand packets which allow to control many aspects during a sailboat navigation mission
`xb_local`	subscription	XBee proxy components	Provides periodically to subscribers information about the XBee link. The data provided by `xb_local` packets is in relation mainly to the quality of the signal with the boat (RSSI, XBee packet retries, etc.)
`xb_local_param`	request-response	XBee proxy components	Through this service is possible to change the power behavior (emission power) of the XBee link between the sailboat and the XBee proxy component involved

Table 2 A-Tirma communication system. ISE components

Components			
Component	Services provided	Services used	Description
XBee proxy component	`telemetry`, `telecommand`, `xb_local` and `xb_local_params`	none	It is a proxy component to communicate with the sailboat through an XBee 868 Pro radio link. Allows other components to get subscribed for telemetry information, and accepts telecommands addressed to the boat by requester components
GPRS proxy component	`telemetry` and `telecommand`	none	Proxy component for communication with the boat through a GPRS data link via TCP/UDP protocols. Allows other components to get subscribed for telemetry information, and accepts telecommands addressed to the boat by requester components
Iridium SBD proxy component	`telemetry` and `telecommand`	none	Proxy component which communicates with the sailboat through the Iridium Short Burst Data (SDB) service via HTTP and email protocols. Allows other components to get subscribed for telemetry information, and accepts telecommands addressed to the boat by requester components

(continued)

Table 2 (continued)

Components			
Component	Services provided	Services used	Description
GUI client component	none	`telemetry`, `telecommand`, `xb_local` and `xb_local_params`	This component does not provide any ISE service, and can use any of the services provided by XBee, GPRS and Iridium SBD proxy components. It is a subscriber and requester of telemetry and telecommand (and `xb_local` and `xb_local_param`) services, respectively
Web server component	none	`telemetry`	This component is a subscriber of telemetry services, and provides a web page service with A-Tirma's tracking and telemetry information during a mission
Name server	ISE naming services		Component which provides ISE naming services. Generic component provided by the middleware
Instantiation server	ISE remote instantiation services		This is a generic component provided by the middleware itself. It provides ISE remote instantiation services

the robustness of the embedded autonomous system on-board and of the distributed communication software infrastructure built using ISE. Concretely, in this 10-hour stress test, we deployed the system using a *short range comunication* node, and two *communication* nodes running remotely in different machines in our laboratory, where one of them was just a secondary node for backing the main one in case it failed. As to *user* nodes, we deployed on-field two nodes running the GUI component of Fig. 10 in two different notebooks under GNU/Linux.

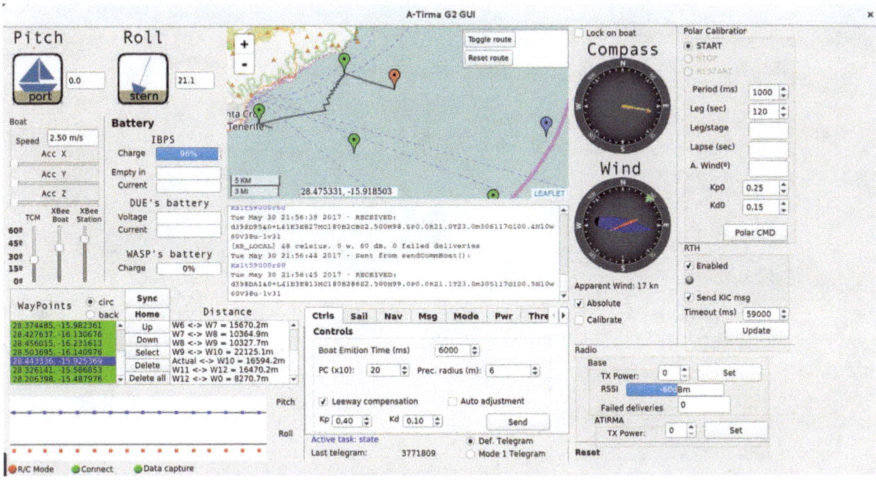

Fig. 10 A-Tirma's control and monitoring interface

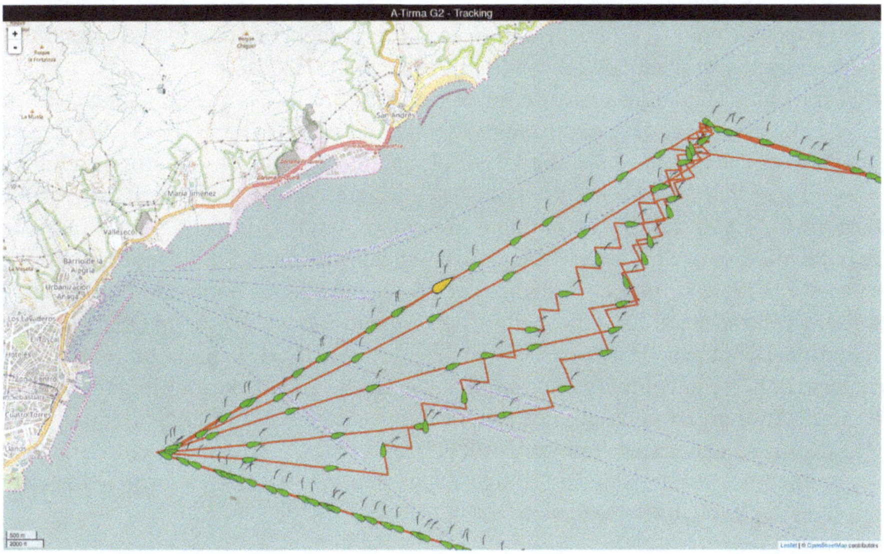

Fig. 11 A-Tirma's web tracking web page

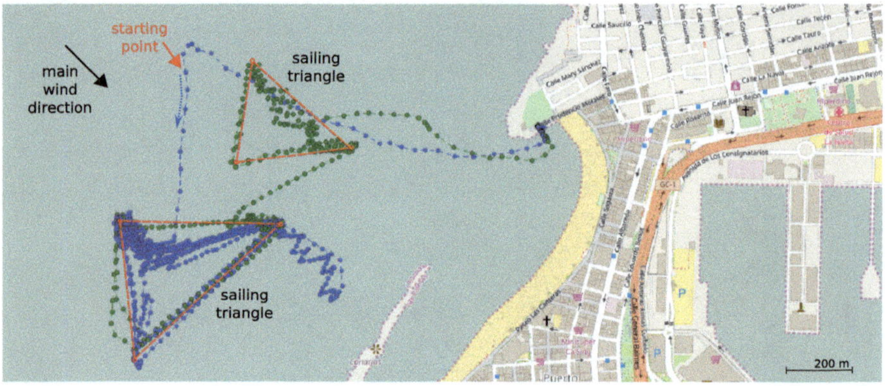

Fig. 12 A-Tirma's 10-hour stress test tracking log

4 Conclusions

In this paper we have outlined the design and development of a distributed component based and service-oriented C++ middleware called ISE, designed and developed to build the communication software infrastructure of an oceanic autonomous sailing vessel called *A-Tirma*. One of the main objectives of developing ISE was to try to take advantage of the new C++ standard, C++11, and its upgrade, C++14, for developing software for distributed complex systems. Furthermore, we have used exclusively standard C++ and the highly portable and quasi standard Boost C++ Libraries to keep the middleware portable, taking advantage of the important metaprogramming features available in this new standards. As an interesting point of simplification of use, we have made an important effort to avoid the use of an specific Interface Definition Language (IDL), in ISE the IDL is C++ itself. This is quite common in other tools for developing distributed software, as CORBA [22], in general, and ROS [23] in the particular area of robotics, as paradigmatic examples. Finally, we have deployed and tested this infrastructure in a real experimental setup to assess its continuous operation during a significant interval of time in terms of system robustness and reliability with satisfactory results.

Acknowledgements The authors are sincerely grateful to Solumatica Canarias for providing financial support for building the A-TIRMA G2 prototype, and to the Real Club Náutico de Gran Canaria and to the Real Club Victoria for the access they granted to their facilities, and for the logistical support during the development of this project.

References

1. Vale, T., I. Crnkovic, E.S. de Almeida, P.A. da Mota Silveira Neto, Y.C. Cavalcanti, and S.R. de Lemos Meira. 2016. Twenty-eight years of component-based software engineering. *Journal of Systems and Software*, 111:128–148. http://www.sciencedirect.com/science/article/pii/S0164121215002095.
2. Oliveira, L.B.R., F.S. Osório, and E.Y. Nakagawa. 2013. An investigation into the development of service-oriented robotic systems. In *Proceedings of the 28th annual ACM symposium on applied computing*, 223–228, Ser. SAC'13. New York, NY, USA: ACM. https://doi.org/10.1145/2480362.2480410.
3. Coulouris, G.J.D.T.K.G.B. 2011. *Distributed Systems: Concepts and Design*, 5th edn. Boston, MA, USA: Addison-Wesley Longman Publishing Co., Inc.
4. Domínguez-Brito, A.C., B. Valle-Fernández, J. Cabrera-Gámez, A. Ramos-de Miguel, J.C. García. 2016. *A-TIRMA G2: An Oceanic Autonomous Sailboat*.Cham: Springer International Publishing, ch. A-TIRMA G2: An Oceanic Autonomous Sailboat, pp. 3–13. https://doi.org/10.1007/978-3-319-23335-2_1
5. Stroustrup, B. 2005. *Abstraction and the C++ Machine Model*, 1–13. Berlin: Springer. https://doi.org/10.1007/11535409_1.
6. Isocpp.org. 2017. C++ Super FAQ, When will compilers implement C++14?. https://isocpp.org/wiki/faq/cpp14#cpp14-compilers.
7. Stroustrup, B. 2013. *The C++ Programming Language*, 4th edn. Boston: Addison-Wesley Professional. http://www.stroustrup.com/4th.html
8. Boost.org. 2017. Boost C++ Libraries. http://www.boost.org
9. Brugali, D., and P. Scandurra. 2009. Component-based robotic engineering (part i) [tutorial]. *IEEE Robotics Automation Magazine* 16 (4): 84–96.
10. Brugali, D., and A. Shakhimardanov. 2010. Component-based robotic engineering (part ii). *IEEE Robotics Automation Magazine* 17 (1): 100–112.
11. Ellis, C., and S. Gibbs. 1989. Active objects: Realities and possibilities. In *Object-Oriented Concepts, Databases, and Applications*. New York: ACM Press, Addison-Wesley.
12. Kohlhoff, C. 2017. Boost. Asio. Boost C++ Libraries. http://www.boost.org/doc/libs/release/libs/asio/
13. Ramey, R. 2017. The boost serialization library. In *The Boost C++ Libraries*. http://www.boost.org/doc/libs/release/libs/serialization/
14. Karvonen, V., and P. Mensonides. The boost preprocessor library. In *The Boost C++ Libraries*.
15. Morgenstern, K.D. 2017. The boost process library. In *The Boost C++ Libraries*. http://www.boost.org/doc/libs/release/libs/process/
16. Henney, K. 2017. The Boost. Any Library. Boost C++ Libraries. http://www.boost.org/doc/libs/release/libs/any
17. Digi. (2017) XBee 868 Pro specifications. http://www.digi.com/products/wireless-wired-embedded-solutions/zigbee-rf-modules/point-multipoint-rfmodules/xbee-pro-868#specs
18. Giovanni Di Sirio. (2017) ChibiOS free embedded RTOS. http://www.chibios.org
19. Cabrera-Gámez, J., A. Ramos-de Miguel, A.C. Domínguez-Brito, J.D. Hernández-Sosa, J. Isern-González, and L. Adler. 2014. A real-time sailboat controller based on chibios. In *Proceedings of the 7th international robotic sailing conference. Robotic sailing 2014*, eds. F. Morgan and D. Tynan, 77–85, Springer International Publishing. https://doi.org/10.1007/978-3-319-10076-0_7.
20. Libelium. 2017. Waspmote—open source sensor node for the internet of things. http://www.libelium.com/products/waspmote.
21. Cabrera-Gámez, J., A. Ramos-de Miguel, A. Domínguez-Brito, J. Hernández-Sosa, J. Isern-González and E. Fernández-Perdomo. 2013. An embedded low-power control system for autonomous sailboats. In *Proceedings of the 6th international robotic sailing conference. Robotic sailing*, eds. Bars, F.L., and L. Jaulin, 67–79. Springer International Publishing. https://doi.org/10.1007/978-3-319-02276-5_6.

22. Henning, M., and S. Vinoski. 1999. *Advanced CORBA Programming with C++, Ser*. Addison-Wesley Professional Computing Series: Addison-Wesley.
23. Quigley, M., K. Conley, B. Gerkey, J. Faust, T. Foote, J. Leibs, R. Wheeler, and A.Y. Ng. 2009. Ros: an open-source robot operating system. In *ICRA workshop on open source software*, vol. 3, no. 3.2, 5. http://www.willowgarage.com/sites/default/files/icraoss09-ROS.pdf

Part II
Navigation and Anti-collision Algorithms

Voter Based Control System for Collision Avoidance and Sailboat Navigation

Jordan Less'ard-Springett, Anna Friebe and Maël Le Gallic

Abstract Autonomous sailboat control systems quickly grow in complexity with a variety of factors needed to be taken into account when deciding a course to sail. These factors include the wind condition, the planned route and marine traffic. This paper proposes a voting based mechanism for deciding the appropriate course. The strategy reduces software complexity while combining navigation within a channel with reactive collision avoidance.

1 Introduction

Autonomous sailing boats offers the potential of a cost-efficient and environmentally friendly Autonomous Surface Vessel (ASV) for marine research. Unmanned wind propelled vessels, due to their low energy requirements, allow for the collection of oceanographic data at high temporal and spatial resolution, during long-term and long-range missions in hazardous conditions. Robotic sailing vessels can also be used for surveillance and transportation of certain goods.

The design of an autonomous sailing boat control system is complex as it has to deal with ASV dynamics, uncertainties, changing winds and sea conditions, stationary and dynamical obstacles, computational issues and control objectives (spatiotemporal, energy, danger level or mission). Wind propelled boat control strategies differ to motorized ASV as they can not navigate in any direction directly, depending on wind conditions and often lack the same speed and maneuverability. On sea, the risk of collision with other vessels and floating objects, such as buoys and sea debris, represents a significant menace. Including a reliable and fast collision avoidance sys-

J. Less'ard-Springett (✉)
Aberystywth University, Aberystwyth, Wales, UK
e-mail: jos60@aber.ac.uk

A. Friebe · M. Le Gallic
Åland University Of Applied Sciences, PB 1010, 22111 Mariehamn, Åland, Finland
e-mail: anna.friebe@ha.ax

M. Le Gallic
e-mail: mael.legallic@ha.ax

K. I. Øvergård (ed.), *Robotic Sailing 2017*,
https://doi.org/10.1007/978-3-319-72739-4_5

tem in the control system is thus a priority in order to navigate safely and efficiently around these obstacles.

In this paper, a voter based system for reactive collision avoidance and sailboat navigation is described. The paper's focus is the modular software architecture that allows for integration of these two components. The proposed architecture is restricted to a previously planned route defined by a set of waypoints. The ASV is allowed to deviate somewhat from the lines connecting the waypoints, in order to handle differing wind conditions and avoid obstacles. The proposed system does not re-plan the route and move waypoints, thus there is need for a complementing higher level long range collision avoidance system. A basic outline of the resulting hybrid system will be described in Sect. 1.2. Also, there is no formal guarantee that avoidable collisions are always prevented by the proposed system. It has been developed for the control of the ASPire (Autonomous Sailing Platform), see Sect. 1.2.

1.1 Related Work

An overview of collision avoidance for autonomous surface vessels can be found in [1]. Collision avoidance includes:

1. Global path planning for avoidance of static obstacles, deliberate methods that can be conducted offline;
2. Path re-planning, deliberate methods conducted online for long range obstacle avoidance;
3. Reactive techniques for short range obstacle avoidance.

A hybrid structure of deliberate and reactive techniques is generally required to achieve a sufficiently well working obstacle avoidance system. The system described in this paper falls into category (3). This is described in Sect. 2.2.

Voter-based systems for obstacle avoidance have previously been described in unmanned surface vehicles [2], and trace back to work at Carnegie Mellon University [3–5] for Lunar rovers among others.

Methods developed in [6, 7] calculate a cost for possible courses as a sum of several different cost functions. The course with the smallest cost is selected. These methods are similar to voter-based methods.

The International Regulations for Preventing Collisions at Sea, COLREGS [8] for short, lay out navigation rules for ships. These rules were designed for humans, and the how these rules are interpreted is not often the same and it is often down to the ships captains experience and common sense to prevent a collision. This is discussed heavily in [9].

There are several proposals for automating COLREGS [10, 11]. In this paper, COLREGs are not considered, but a discussion is provided in Section Conclusions and Future work.

1.2 ASPire Project

The ASPire is a wind propelled ASV developed by the Åland University of Applied Sciences that will be deployed for collecting marine sensor data (i.e., Conductivity, Temperature and Density (CTD) measurements) and for passive-acoustic monitoring (PAM) of harbor porpoise in the vicinity of the Aland Islands in the Baltic Sea. It is propulsed by a free rotating rigid wingsail, equipped with a single-axis solar tracker for solar energy harvesting and uses a windvane self steering device in addition to an electric actuator for steering the rudder [12].

To detect nearby and distant obstacles, a forward-looking Thermal Imaging Camera (TIC) and automatic identification system (AIS) are used. Due to the low power consumption and size constraints of the ASPire, the selected sensors are:

- A FLIR MD-324 Thermal Imager Camera, with a field of view of 24 × 18, for real-time short-range (hundreds of meters) obstacle detection;
- A Class B AIS CTRX Graphene for vessel detection in a range of around 10 NM.

AIS is a tracking system used by marine traffic to avoid collisions at sea and provides the position, course, speed and size of other nearby AIS-equipped vessels. Maritime laws require the use of the AIS on voyaging ships (commercial and passenger ships) with gross tonnage (300 or more) and on European Union fishing boats over 16 m.

Obstacle detection using monocular vision provides an estimation of the bearings from the USV to the detected obstacles.

An hybrid collision avoidance strategy based on these sensors properties was designed for the ASPire. The high level deliberative layer estimates the risk of collision of the ASV navigating in its predefined waypoint route, with all AIS detected vessels. This collision detection is performed for a defined interval of time, by assuming the AIS vessel speeds and courses constant. Then, if a potential collision is detected, the waypoint route is re-planned from the potential collision route section. Using interval analysis for implementing theses two steps, allows to easily take into account sensors data uncertainties and provides a guaranteed collision-free route when the assumptions made are true [13]. The high level collision avoidance main objective is to keep the ASV away from large ships. The low level reactive layer uses the voter based system described in this article for avoiding close range obstacles.

All the code is open source and can be found on the Åland Sailing Robots GitHub repository [14].

2 Voter Based Architecture

A challenge when developing sailboat controllers is handling the different factors that need to be considered when choosing a course for the sailboat to sail. When considered as a single problem it can become unwieldy from a software perspective. Thus the problem was approached by attempting to break it down into smaller discrete components that were focused on a single aspect of controlling the sailboat. Each component gives a value to the possible courses the sailboat could sail. This value reflects how well the course supports the objective of the component. We refer to this value as the number of votes, and the components as voters. The votes for each course are then summed together by a module known as the Arbiter, and the course with the most votes is chosen as the desired course for the sailboat to sail. Voters can also be given a weight, their votes are multiplied by this value. The weight describes the importance of the voter, for example avoiding a collision is more important than heading towards a target waypoint. The maximum votes a voter can place on a single course is controlled by a parameter referred to as $vote_{max}$ throughout this paper.

This architecture provides a modular and loosely coupled design, with small independent voters that each focus on one specific issue, such as the direction to the waypoint. This is an application of the separation of concerns design principle, where the software complexity of each voter can be reduced, compared to considering all the factors at once. In addition, this design provides a flexible framework for allowing multiple sailing configurations, such as traditional sails and wing sails, by swapping in and out different voters, which maximises code reuse and simplifies maintenance. Voters could also be allowed to veto courses which is used often in similar designs [5]. This could be used to prevent the sailboat from navigating a course even if it received the most votes such as preventing the sailboat from sailing into the wind, under any circumstances. However it is worth noting that we are currently not using vetoes.

2.1 Core Sailing Voters

Four voters make up the core sailing logic: Waypoint Voter, Wind Voter, Maneuver Voter and the Channel Voter. The following sections provide a high level overview and purpose of the individual voters. Theta, θ is used to denote the course being voted on.

2.1.1 Waypoint Voter

The Waypoint Voters sole purpose is to ensure the boat is heading towards its target waypoint. This voter places votes on courses that bring it closer to the waypoint. The closer the course is to the bearing of the waypoint from the autonomous sailboat, the more votes it receives.

The waypoint voter will assign its votes as follows:

$$vote_\theta = \begin{cases} vote_{max} \times (1 - \frac{|d|}{90}) & \text{if } |d| < 90 \\ 0 & \text{otherwise} \end{cases} \tag{1}$$

where d is the shortest angular difference between the bearing to waypoint and the possible course, θ. d is in the range $[-180, 180]$.

2.1.2 Wind Voter

The Wind Voter aims to keep the autonomous sailboat from sailing into the no-go zone, the direction from which the wind is coming from, and to encourage the sailboat to sail an effective tack course, at the edge of the no-go zone when necessary. The voter does this by assigning a number of votes, two thirds of the maximum vote was found to be a good number; to every course apart from those in the no-go zone, which receive 0 votes. This encourages the sailboat to not sail into the no-go zone. To encourage effective tacking, when the bearing to the waypoint is within the no-go zone, additional votes are added to the tack courses. Without these additional votes, when the sailboat needs to tack, its tacks are shorter, and so the sailboat is slower.

Where t is the true wind direction, angles are assumed to wrap around and the shortest angular difference is used.

$$vote_{\theta} = \begin{cases} 0 & t - 45 >= \theta < t + 45 \\ vote_{max} \times \frac{2}{3} & otherwise \end{cases} \tag{2}$$

During simulation experiments as the ASV moved, and courses were evaluated, a new course would often be chosen, winning out by a vote or two. This leads to the ASV not sailing a smooth path, and would often cost it speed. Thus a small additional vote was applied to the course the ASV was already sailing so the ASV would reduce the amount of minor course changes it would make which would hurt the ASVs overall performance. This additional vote could have been applied by a separate voter, similar to the maneuver cost function applied in [7].

2.1.3 Maneuver Voter

The maneuver voter was created in response to several early simulation experiments. As the sailboat moved, and courses were evaluated, a new course would often be chosen, winning out by a vote or two. This leads to the sailboat not sailing a smooth path, and would often cost it speed.

This voter places a number of votes in the direction the sailboat is already sailing. This reduces the amount of minor course changes that occur which hurt the sailboat's overall performance.

2.1.4 Channel Voter

While a ASV is navigating a waypoint, there is huge amount of area it could sail to before hitting the waypoint. Being able to constrain the sailboats path to a sailing channel between two waypoints gives us more control of where the sailboat goes. This allows the sailboat to navigate in narrow bodies of water, improving its operational areas. It also provides a good mechanism for deciding when to change tack and control how far the sailboat can travel on a single tack.

The Channel Voter paints an imaginary line between the last waypoint and the current waypoint, the sailboat s signed distance to this line is used by the voter to determine how close to the channel edge it is, and thus how it should vote. When the sailboat is not near the edge of the channel, the Channel Voter places 0 votes across every course. When the sailboat approaches the edge of the channel, the voter pushes the sailboat back towards the centre line.

$$vote_{\theta} = \begin{cases} vote_{max} & |b| < 45 \\ 0 & otherwise \end{cases} \tag{3}$$

where b is the shortest angular difference between the course θ and the perpendicular course towards the centre line.

2.2 Collision Avoidance

Due to the two different detection systems available on Aspire, AIS and the thermal imaging system, the problem of collision avoidance was decomposed into two voters, one for dealing with close range contacts(Proximity Voter), and one for dealing with contacts further away (Mid Range Voter). This also allows us to use different strategies for contacts that are close and pose more risk, and contacts that are further away from that pose less risk.

2.2.1 Proximity Voter

The Proximity Voter is one of the two collision avoidance voters. Its purpose is to prevent an immediate collision from nearby AIS contacts, as well as contacts picked up by the thermal imaging system. If there are no nearby contacts, then the proximity voter doesnt place any votes. However if there are contacts, the proximity voter places votes on courses that take the sailboat away from the detected contacts, and their courses where that data is available.

The voter uses different approaches based on the type of contact it is processing due to the different data they offer.

When avoiding an AIS contact, the Proximity Voter examines the course of the contact, and places votes that are perpendicular to this course which are on the same side of the contacts course as the sailboat. The other side receives negative votes, so that any existing votes that have been placed here due to other contacts are not considered as highly due to the chance of a collision. As well as votes being place on the perpendicular course, votes are also placed either side of these courses, decreasing as they move away from the perpendicular course. Only courses 40° either side were given votes, this is to ensure the sailboat is always moving away from the contact.

$$vote_\theta = \begin{cases} vote_{max} \times \frac{D}{s} & if|p| < 40 \\ 0 & otherwise \end{cases} \tag{4}$$

where D is a tuning parameter, which controls the minimum safe distance and s is the distance to the contact. p is the shortest angular difference between θ and the perpendicular course away from the contact. Where the value of $vote_\theta$ is clamped to $vote_{max}$

When the voter is avoiding a visual contact that has been picked up the thermal imaging system the only data we have available is the bearing to the target. The approach we use to avoid a visual target is rather simple: the courses that bring us towards the visual contact receive negative votes, and the courses that take us away receive positive votes. Just like with the AIS contacts, the votes decrease/increase linearly as you get further away from the contacts bearing.

$$vote_\theta = \begin{cases} vote_{max} \times \frac{D}{s} & if|p| < 40 \\ 0 & otherwise \end{cases} \tag{5}$$

where d is the shortest angular difference between θ and the bearing away from the contact, the inverse of the bearing to the contact.

The Proximity Voter is a weighted voter, this means that its votes are multiplied by a factor, we used the value 1.5. This weighting means its votes are worth more than the other voters to reflect that avoiding a collision is more important than the other voters goals.

2.2.2 Mid-Range Voter

This is the second of the two collision avoidance voters. Instead of avoiding nearby contacts, it tries to keep the sailboat away from AIS contacts that are further away assuming that near collision wont happen. It does this by examining how a course the sailboat could take intercepts the course of a contact, and determining the risk of collision. Initially every course receives the maximum vote the voter can give, and then this value is reduced based on collision risks, with a course that has a high risk of collision receiving no votes.

The Mid-range Voter uses a similar approach to the one laid out in [15]. It examines the closest point of approach, and uses this to estimate a collision risk. Courses that have a close closest point of approach (100 m or less is deemed as the highest collision risk for the duration of this project) receive less votes than a course whose closest point of approach is higher (1 km being the max). The time to approach is not been a consideration in this project, as finding the right value for this will require significant testing, and given the limited time available this was not done. However a more complete solution would certainly want to take this into account.

The voter works by iterating over all the possible courses, and then calculating the closest point of approach of that course with every AIS contact. The contact with the lowest closest point of approach is stored, and once every contact has been

examined, the lowest closest point of approach is used when calculating the number of votes to place on this course. This approach allows every contact to be considered for every course the sailboat can take. The amount of votes given is based on how close the closest point of the approach is compared to the safe distance. The formula is shown below
Where m is the minimum safe distance and where c is the closest point of approach.

$$vote_\theta = \begin{cases} vote_{max} \times \frac{m-c}{m \times vote_{max}} & if|p| < 40 \\ 0 & otherwise \end{cases} \tag{6}$$

3 Results

The voter based system previously described was simulated in a variety of different configuration. For each configuration set two simulations were performed, one with the collisions avoidance voters and the other without. The Table 1 shows off the minimum distance of the closest point of approach (MDCPA) between the ASV and the contact vessel for all those simulations. The test name refers to the contacts path in

Table 1 Simulation results, CAV refers to collision avoidance voters

Simulation configurations				MDCPA	
Simulation	Down wind	Initial distance	Contact speed	**With CAV**	**Without CAV**
1. Head on	Y	500 m	3 m/s	98.7 m	14 m
2. Head on	N	500 m	3 m/s	94 m	57.6 m
3. Starboard crossing	Y	500 m	3 m/s	101.28 m	63.7 m
4. Starboard crossing	N	500 m	3 m/s	135.8 m	131.2 m
5. Port cross	N	500 m	3 m/s	187.94 m	193.57 m
6. Port cross	Y	500 m	3 m/s	97.5 m	63.7 m
7. Head On	Y	1 km	10 m/s	92.6	16.7 m
8. Head On	N	1 km	10 m/s	79 m	83 m
9. Take Over	Y	1 km	10 m/s	68 m	10.2 m
10. Take Over	N	1 km	10 m/s	80.5 m	79.7 m
11. Head On & Take over	Y	1 km	10 m/s	92.9 m	19.5 m
12. Head On & Take over	N	1 km	10 m/s	82.3 m	83.2 m
13. Head On and cross	Y	500 m & 1 km	3 m/s & 10 m/s	79.9 m	19.5 m

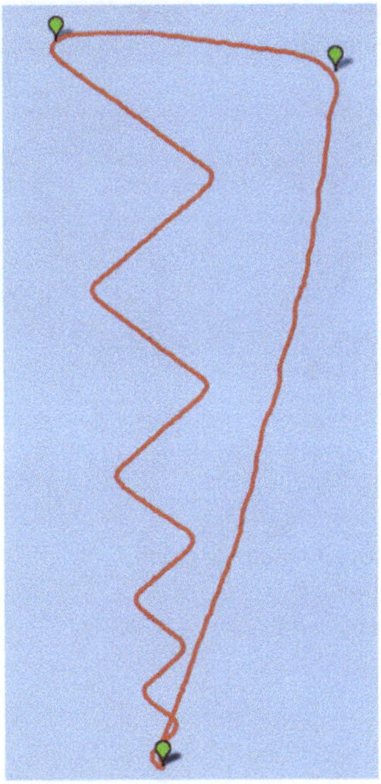

Fig. 1 Core voters simulated

relation to the sailboat. It displays that the control system is able to greatly increase the CPA compared to when the simulation is run without the collision avoidance voters. When there is no chance of collision it does not induce a chance of collision.

Two speeds of the contacts were chosen for the tests: 3 m/s, which is similar to the traveling speed the ASV in simulation and 10 m/s, which is similar to the speed of a large ship which the sailboat might encounter. The final three tests listed in the table have two contacts simulated, one that is "overtaking" the sailboat, and one that is on a collision course, and the final test has a contract crossing the path of the sailboat instead of "overtaking".

Figure 1 shows the simulated path (coloured red) of a sailboat running the core sailing voter. The green balloons are the three waypoints, with the course going counter clockwise. The wind was coming from the south. The sailboat started at the bottom most waypoint, with its first target waypoint being the top right waypoint, a downwind leg. The sailboat completes the course with a reasonable path. The upwind leg, shows tacks that get shorter and shorter as it gets closer to the waypoint. This behaviour is due to how the waypoint voter places its votes on courses based on their distance from the bearing to the waypoint. The black dotted lines in Figs. 2 and 3 are the sailing channel boundaries.

The following two figures show a simulation of the sailboat again (path coloured red) but with the collision voters also running. These experiments have simulated marine traffic in them, the paths of the simulated traffic are also shown. Their paths

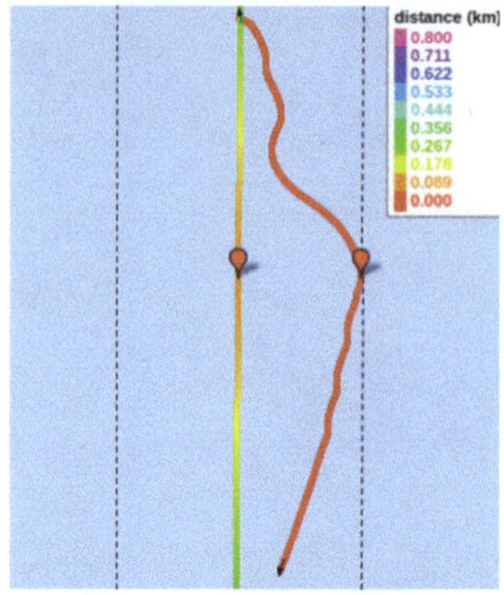

Fig. 2 Single contact simulated

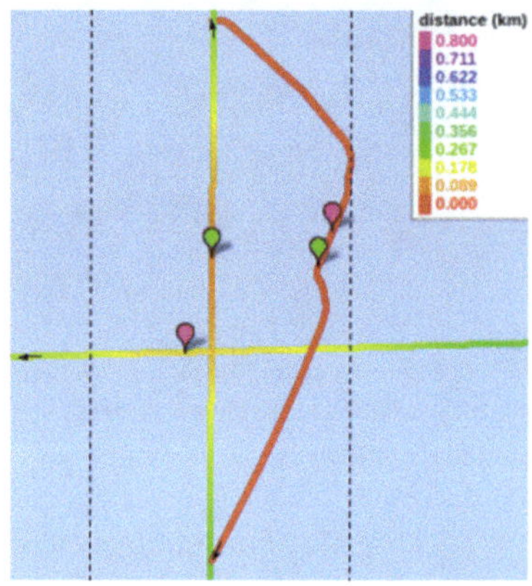

Fig. 3 Multiple contacts simulated

change colors based on the distance to the simulated sailboat, this was done so that the distance across the whole simulation can be seen.

Figure 2 shows a single contact sailing north, this is the path of simulation 1 in Table 1. Here you can see the sailboat avoiding the contact on its starboard side before returning back to its tack, and navigating to the waypoint. The red balloons indicate the positions of the contact and the ASV when they were at their closest. The sailboat can been seen just pushing past the sailing channel due to the weighting of the Proximity Voters votes.

The simulation labelled 13 in Table 1 is displayed in Fig. 3. There are 2 contacts: one contact is traveling from south to north at 10 m/s and the other is traveling from east to west at 3 m/s. The balloons indicate the positions of the sailboat and the contacts at their closest point. The green pair of balloons indicates the position of one contact and the ASV at their closest while the red pair indicates the position of the other contact and the ASV at their closest.

4 Conclusion and Future Work

The voter based control system is a well suited, modular architecture that allows for further development of control algorithms as well as collision avoidance strategies. Simulations give encouraging results, and imply that the collision avoidance algorithms are useful for obstacles in the short to mid range distances, up to 1–2 nm. The modular architecture simplifies the maintenance and reusability of the Åland Sailing Robots codebase which has to support a variety of different type of autonomous sailboats.

Future work will involve testing of the voter based system on the water in real world conditions with the core sailing voters before testing the collision avoidance voters. Tuning of the voters' parameters will also take place. A wing sail voter also needs to be developed for the ASPire which behaves differently to the traditional sail the current wind voter is based around.

Regarding COLREGs we see two situations. For large ships, there is no guarantee the ship will make an avoidance maneuver when the ASV is deemed as being unmanned, and unable to harm the ship. Therefore a defensive approach to keep the ASV at a safe distance is deemed suitable. This will be handled by the higher level deliberate replanning, and to some extent by the mid-range voter if necessary.

In a situation where the autonomous vessel is close to another vessel it is reasonable to attempt to automate COLREGs. The behavior described in COLREGs is what the manned vessel would expect. To truly automate COLREGs is challenging, as the autonomous vessel needs to interpret whether the manned vessel is for example sailing or fishing to evaluate which of the vessels is the stand-on versus give-way vessel.

The Proximity Voter can be changed to comply with COLREGs in a simplified manner, such as implementing rule 19 (d) for how to conduct in limited visibility. This would mean that the sailboat would avoid all close vessels by a starboard maneuver, except if the case is a vessel approaching from behind to perpendicular on starboard side. In this case the sailboat would perform a maneuver to port side.

One particular voter that will need some future changes is the Channel Voter. Being able to designate a particular part of the route where the sailboat should never leave the sailing channel is a crucial development. This is needed before the sailboat could be left unattended in certain areas, such as those close to the shore. This paper presents a modular system suitable for future exploration of control algorithms and collision avoidance strategies.

Acknowledgements ASPire is part of the marine research project which is partly funded by the European Regional Development Fund [16].

EUROPEAN UNION
European Regional Development Fund

References

1. Campbell S, Naeem W, Irwin G W. 2012. A review on improving the autonomy of unmanned surface vehicles through intelligent collision avoidance manoeuvres. *Annual Reviews in Control*, 36(2): 267–283, Elsevier.
2. Larson, J., M. Bruch, R. Halterman, J. Rogers, and R. Webster. 2007. *Advances in autonomous obstacle avoidance for unmanned surface vehicles*. Report: Defense Technical Information Center.
3. Simmons R, Henriksen L, Chrisman L, Whelan G. 1998. Obstacle avoidance and safeguarding for a lunar rover. In *Proceedings from the AIAA forum on advanced developments in space robotics*, Madison, WI.
4. Rosenblatt J K. 1997. DAMN: A distributed architecture for mobile navigation. *Journal of Experimental and Artificial Intelligence*, 9(2–3): 339–360. Taylor & Francis.
5. Singh S, Simmons R, Smith T, Stentz A, Verma V, Yahja A, Schwehr K. 2000. Recent progress in local and global traversability for planetary rovers. In *Proceedings of IEEE international conference on robotics and automation (ICRA 2000)*, Vol. 2, pp. 1194–1200.
6. Guo Y, Romero M, Leng S, Plumet F, Benosman R, Gas B. 2011. Reactive path planning for autonomous sailboat using an omni-directional camera for obstacle detection. In *IEEE international conference on mechatronics*.
7. Wirz J, Tranzatto M, Liniger A, Colombino M, Hesse H, Grammatico S. 2016. AEOLUS, the ETH autonomous model sailboat. In *Robotic sailing 2015*, pp 103–112. Springer International Publishing.
8. International Maritime Organization: Convention on the International Regulations for Preventing Collisions at Sea, 1972 (COLREGs). http://www.imo.org/en/About/conventions/listofconventions/pages/colreg.aspx. Accessed 30 May 2017.
9. Benjamin M R, Curcio J A. 2004. Colregs-based navigation of autonomous marine vehicles. In *Autonomous underwater vehicles, IEEE/OES*.
10. Statheros, T., G. Howells, and K.M. Maier. 2008. Autonomous ship collision avoidance navigation concepts, technologies and techniques. *The Journal of Navigation* 61 (01): 129–142.
11. Perera L P, Carvalho J P, Soares C G. 2008. Autonomous guidance and navigation based on COLREGs rules and regulations of collision avoidance. In *Proceedings of the international workshop advanced ship design for pollution prevention*, pp 205–216, Taylor and Francis Group, London.
12. Friebe A, Olsson M, Le Gallic M, Less'ard Springett J, Dahl K, Waller M. 2017. A marine research ASV utilizing wind and solar power. In *Proceedings of Oceans17 MTS/IEEE aberdeen*.
13. Jaulin Luc, Le Bars Fabrice. 2012. *A simple controller for line following of sailboats. Robotic sailing 2012*, 107-118. Springer.
14. Åland Sailing Robots Codebase. https://github.com/AlandSailingRobots/sailingrobot/. Accessed 30 May 2017.
15. Alves J, Cruz N. 2016. AIS-Enabled collision avoidance strategies for autonomous sailboats. In *Robotic Sailing 2015: proceedings of the 8th international robotic sailing conference*, pp 77–87. Springer.
16. European Union Regional Development Fund. http://ec.europa.eu/regionalpolicy/en/funding/erdf/. Accessed 30 May 2017.

An Attractor/Repellor Approach to Autonomous Sailboat Navigation

Dermot Tynan

Abstract This paper proposes an alternative navigation method which avoids the use of traditional waypoints and cross-track error, by utilising an array of waypoint attractors and repellors of different strengths and polarities. A range of possible headings either side of the bearing to the waypoint are considered, and the velocity-made good on each heading is computed. The heading with the optimal VMG is chosen. Attractors are waypoints which have a positive effect on the route (like traditional waypoints) and repellors are waypoints which have a negative or repelling effect on the heading.

1 Introduction

Sailboats, particularly those involved in offshore racing, normally have a well-defined demarcation between helming and navigation [8]. Helming a boat at sea is a continuous operation requiring a certain skill-set. It involves responding to changes in wind direction and strength, heeling angle, leeway, course and other variables, and adjusting the rudder to maintain a constant heading in relation to the compass or to the apparent wind angle [7]. Changes in the apparent wind or the compass heading are corrected by adjusting the rudder angle until the correct wind angle or compass heading is achieved. The basics of this function can be replaced by a PID controller [1, 12, 18] or an Autohelm [14].[1]

In contrast, sailboat navigation looks at the yacht position from the GPS, celestial sighting or from dead-reckoning, and determines if the current course is ideally suited to reaching the destination. Navigation will also factor in weather forecasts, obstructions, hazards and other operational details to determine the best heading. In general terms, a navigator will choose a set of waypoints (known as a route) and determine the best course from waypoint to waypoint [4, 10]. As sailboats cannot

[1] Autohelm is a trademark of Raymarine.

D. Tynan (✉)
Kalopa Research, Aughinish, Ireland
e-mail: dtynan@kalopa.com

K. I. Øvergård (ed.), *Robotic Sailing 2017*,
https://doi.org/10.1007/978-3-319-72739-4_6

make progress upwind, the course may involve tacking back and forth to make the waypoint. Also, tidal currents can push the vessel to one side or other and force the navigator to add a tidal correction. Navigating is not a full-time activity, however. It is possible to produce a set of waypoints and to review the course on a periodic basis to determine if the optimal route is being achieved [4, 8]).

The partition of responsibilities between helmsman and navigator allows for a subsumption architecture [6, 17] in a robotic sailboat. A low-level processor which is always running, such as an Arduino, can measure the apparent wind angle or the compass course and compute the error against the required angle, and adjust the rudder to bring the error to 0. The low-level system need not understand latitude or longitude, great-circle distances, or other floating-point complexities. Instead, it is little more than an autopilot or autohelm, possibly with some additional capabilities such as environmental monitoring and alarm detection [2, 3, 16, 18].

An upper-level system with access to a GPS for time and position information, as well as logging and satellite communications facilities, can handle the navigation tasks on an as-needed basis. As this function is not real-time in the same way as the low-level system but involves more advanced floating-point calculations, it is possible to use a more advanced processor and to power it up only when necessary. This approach can introduce significant time-series anomalies in the GPS data, however. It is not entirely dissimilar to severe tidal effects or intermittent GPS reception, both of which can also present complex challenges to robotic sailboats.

The cross-track error (or XTE) is defined as the distance from the vessel's present position to the closest point on a line between the origin and destination waypoints of the leg of the active route [13]. The navigator must allow for a reasonable maximum XTE when choosing waypoints and routes around hazards such as headlands [7]. The chosen route must allow for a maximum XTE which keeps the vessel free from obstructions and hazards [4]. Where GPS data is incomplete, or the navigation

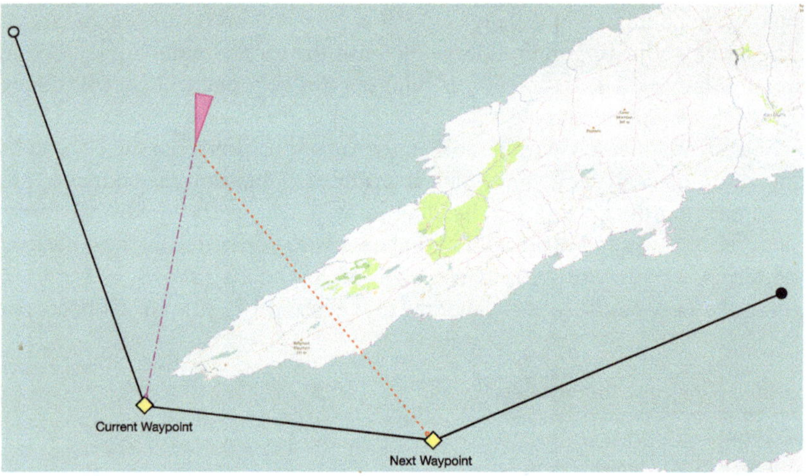

Fig. 1 Hazards of autonomous navigation where Cross-Track Error (XTE) is violated

system is offline for significant periods, it is possible for the boat to exceed the planned maximum XTE. Likewise, strong tidal currents can also push the vessel into unplanned or uncharted territory. As a result, the optimal course to make the next waypoint may in fact drive the vessel onto a headland (see Fig. 1).

2 Navigation by Attractors and Repellors

This paper proposes an alternative navigation method which avoids the use of traditional waypoints and cross-track error, by utilising an array of positive and negative waypoints, referred to here as attractors and repellors. Each of which has a different strength and polarity, much in the same way as a magnetic field may attract or repel certain magnetic objects. A normal waypoint is comprised of a latitude/longitude tuple which uniquely defines a specific position on the surface of the planet. However, due to GPS location inaccuracy and other errors, a waypoint is typically considered with a radius or minimum distance. To achieve a waypoint on a given route, it can suffice to be less than a nautical mile away from the waypoint. In some cases, such as in the middle of an ocean, the acceptable distance to the waypoint might be as much as ten or more nautical miles. In a narrow channel, the distance may be much less than a mile.

Navigation normally proceeds by computing the distance and bearing to the current waypoint. If the distance is less than the minimum, the waypoint is achieved and the vessel advances to the next waypoint. Otherwise, the bearing becomes the new heading (if permissible) and the boat continues. Even at that, a human navigator will make a conscious decision to declare a waypoint reached if the next waypoint appears favourably close. For example, if the distance to the next waypoint is less than the combined distance to the current waypoint plus the distance from the current to the next one, then it is normally reasonable to assume that the waypoint has been reached.

Again, this assumes that there are no hazards or obstructions which might affect such a decision. Some autonomous boats attempting ocean crossings have been known to hunt for the current waypoint even though it is obvious to a human observer that it is better to advance to the next waypoint, for the aforementioned reason. This is particularly true of motorboats which operate only during daylight hours.

An attractor or repellor waypoint is comprised of the same latitude and longitude tuple, but also includes a minimum distance and a normal angle. The distance and angle describe an arc of 180° around the centre of the waypoint (see Fig. 2), beginning at the *normal angle* and continuing for (*angle* + *180°*). For our purposes, we refer to these as attractors. The vessel is considered to be *0* m away from the attractor if the vessel is inside the area covered by the arc, and in such a case, the attractor is considered to be reached. Normally, the attractors would be arranged in such a way as to bisect the bearings from the previous and the next waypoints.

In and of itself, this does little to improve the navigational integrity of the vessel, apart from allowing a minimum distance per waypoint. However, using an attractor

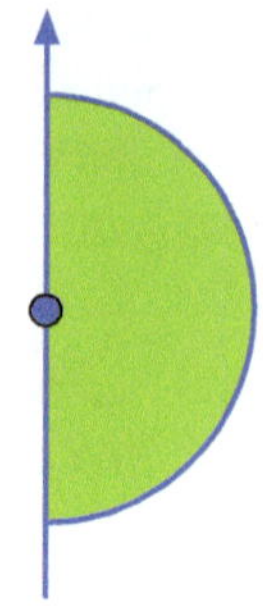

Fig. 2 Attractor (with normal at 0°)

approach, the distance and bearing to all the future waypoints are considered and weighted by their distance or attractor effect. Those farther away, have less influence compared to those close at hand. At the same time, the current attractor cannot lead the vessel in a different direction to the weighted sum of the remaining attractors. This can address the case where the boat has sailed or drifted beyond the waypoint and should not turn around in order to satisfy the minimum-distance criterion. To address the issue of intervening hazards of which the vessel is unaware, the concept of a repellor is introduced. Like the attractor, it is comprised of a latitude and longitude, as well as a normal angle and a distance radius.

However, in terms of navigation, its effect on the proposed course is to repel the boat away from itself. A repellor might be defined on a particular coastline with a minimum distance which covers a reasonable length of the coastline, and with a normal angle which places the area of the arc behind the coastline and covering the hazard itself. See Fig. 3 which shows a protected headland using five repellors.

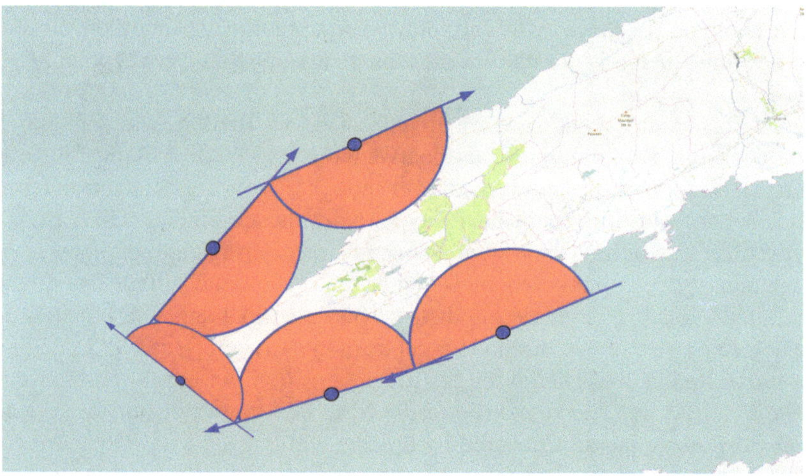

Fig. 3 A headland protected by five repellors

3 Computing Ideal Heading

To consider the mechanism for applying the effects of all the attractors and repellors, it is first important to consider an alternative mechanism for determining the best point of sail for the boat. Many autonomous sailboats attempt to steer the boat by calculating the bearing to the next waypoint, and steering to that heading unless it is within the "no-go zone"; that point of sail which attempts to steer the boat into the wind, or less than 40° of true wind angle [9, 16].

While this addresses the no-go zone appropriately, it does not address improvements which can be achieved by calculating the "velocity made-good" or VMG. VMG is defined as the component of a sailboat's velocity which is acting in the direction of the next waypoint or mark. It can be calculated as:

$$V_{mg} = V_h \cos \alpha \tag{1}$$

where V_{mg} is the velocity made-good, V_h represents the speed at the given heading and α the angle between the desired bearing and the actual heading.

To illustrate this point, some high-performance yachts can reach very high speeds at true wind angles of 125°, and relatively poor performance at true wind angles greater than 170° [15, 19]. As a result, it is better for these boats to gybe or "tack downwind" even though they could sail directly to the waypoint.

Given a theoretical polar curve for the boat, it is possible to compute the VMG for a range of values of α Instead of accepting the bearing as the new heading, we can look for a maximum value of VMG in and around the requested bearing, by examining each of 91 discrete values and choosing the value with the best VMG. Under normal circumstances, α values greater than or equal to 90° will not produce a useful VMG. As a result, we can choose to evaluate headings which are between ±45° either side of the desired bearing. Unfortunately, a side-effect of this mechanism is an over-reliance on tacking in the case where the waypoint is directly upwind. Initially, the best heading might be a starboard tack on a close-hauled point of sail. However, shortly thereafter, the algorithm will determine that a port tack on a close- hauled point of sail is in fact favoured, and the sailboat will tack through the wind, followed almost immediately by a tack back to the previous heading. In this situation, no allowance is made for the cost of the tack, which can slow or even stall the boats upwind progress. To accurately portray the negative benefits of tacking, headings which are on an opposite tack (or gybe) to the current heading, have their polar speed reduced by a specific factor, such as a ten percent speed penalty, which is referred to as the tack bias. As a result, the vessel will only tack through the wind, when the relative improvement in the opposite tack is significant enough to overcome the implied cost of the manoeuvre. Simulations have shown this as a very effective mechanism for computing the best point of sail for a given vessel.

4 Using VMG and Magnetic Waypoints

Having determined that an appropriate algorithm for computing the boats point of sail is to review each possible heading around a desired bearing and compute the best VMG, it is now possible to extend this to incorporate the navigational attractors and repellors which have been defined.

Essentially, a VMG is computed for each of the attractors and repellors, not just the current one. However, the distance to the object is also used to prioritise those which are closer. In this case, we have:

$$VMG(h) = \sum_{n=0}^{N} \frac{V_h S_n}{d_n} \cos(b_n - h) \tag{2}$$

The distance to the waypoint W_n is represented by d_n and b_n is the bearing. S_n is $+1$ for attractors and -1 for repellors.

In this example, we pre-compute V_h which is the predicted hull speed for the proposed heading, and pre-apply the tack bias. It is worth clarifying that $VMG(h)$ in this sense is not a real velocity value, but a weighted sum which is used purely to choose an ideal heading.

Under normal circumstances, it is sufficient to examine proposed headings $\pm 45°$ either side of the bearing to the current attractor. However, in certain circumstances it is advisable to review this arc. For example, in a narrow channel with obstructions either side, the system may consider a wider arc in order to escape the hazards (see the next section). In the system under test, a mechanism is utilised which is a function of the distance to the nearest repellor. If a repellor is within a very small distance then the desired heading is one which is orthogonal to the normal of the repellor, and away from it. In other words, *repellor.normal* $+ 270°$. If instead the repellor is reasonably close, the heading might consider an arc of $\pm 90°$ either side of the current attractor bearing. Where the repellor is not close, the system can consider headings in a more constrained arc of $\pm 45°$ either side of the bearing to the current attractor. These should be tuneable parameters.

By utilising this waypoint mechanism, it is possible for the vessel to lose (and regain) GPS locations or incur significant tidal drift, and end up with a headland between it and the next attractor (as shown originally in Fig. 1). However, due to the presence of the repellors (Fig. 3), it will not choose a course which brings it closer to the hazard, even if that seems to be the most appropriate bearing to the next attractor.

5 Special-Case Navigation

With this protection against situations where navigational short cuts can introduce obstructions into the proposed route, it is possible to optimise the attractor mechanism further. Even though a chosen heading is analysed in terms of its effect on the overall mission, we have not rendered obsolete the notion of a current waypoint. The next attractor in the list is still the basis for our proposed heading computations. As discussed earlier, the attractor is removed from the list once it is achieved. Repellors are never removed.

If the distance to the current attractor, plus the distance between the current and next attractor is greater than or equal to the distance to the next attractor, the current attractor can be considered achieved and removed from the list. This eliminates situations where the vessel has overshot the waypoint and is attempting to address this by sailing back to it. While this is of use in ocean-going missions, it is counter-productive where the course is severely constrained, or where the waypoints are actually marks of the course such as an inshore "Olympic triangle" event at the World Robotic Sailing Championships [21]. There are also cases where the boat may have to navigate in close quarters around a particular headland and the above optimisation could render the topmost attractor redundant. For these reasons, each attractor also includes a Boolean value which determines whether the remaining attractors should be considered. By setting this to TRUE on all attractors, the system functions in a manner similar to traditional mechanisms, but with the addition of repellors for avoiding obstacles.

An interesting problem arises around the selection of waypoints for a start and finish line, particularly that of the Microtransat [5, 11]. Choosing a waypoint on the centre of the start line can cause an autonomous boat to sail down the line to the centre, before starting. Setting a wide radius on the waypoint can result in a boat which is accidentally on the wrong side of the line, incorrectly assuming it has started. Instead setting an attractor a short distance away from the start line, with a distance slightly shorter than half the length of the line, and with a normal which is parallel to the start line but chosen so that the arc is on the pre-start side, followed by a similar attractor on the opposite side of the line, as shown in Fig. 4, will allow the boat to start almost anywhere on the line, cross the line properly, and then proceed to the next real attractor on the course. As discussed in the previous section, as these two attractors are effectively "marks of the course", they should be flagged as described previously so that further attractors on the route are not considered when calculating the ideal heading.

6 Navigation Simulations

Taking an idealised polar curve for a typical sailboat, which can get up to 43° of apparent wind angle to the wind, allows for a comparison of VMG techniques. In Fig. 4, the vessel navigates by computing the most favourable VMG to the current waypoint, while ignoring future waypoints and all repellors. Tack bias in this example is set to 10%, which says that tacking the boat is a high enough speed penalty that predicted VMG values on opposite tacks should be one tenth of those on the current tack. This shows a reasonable "beat" upwind. For the purposes of this simulation, the wind is from the North.

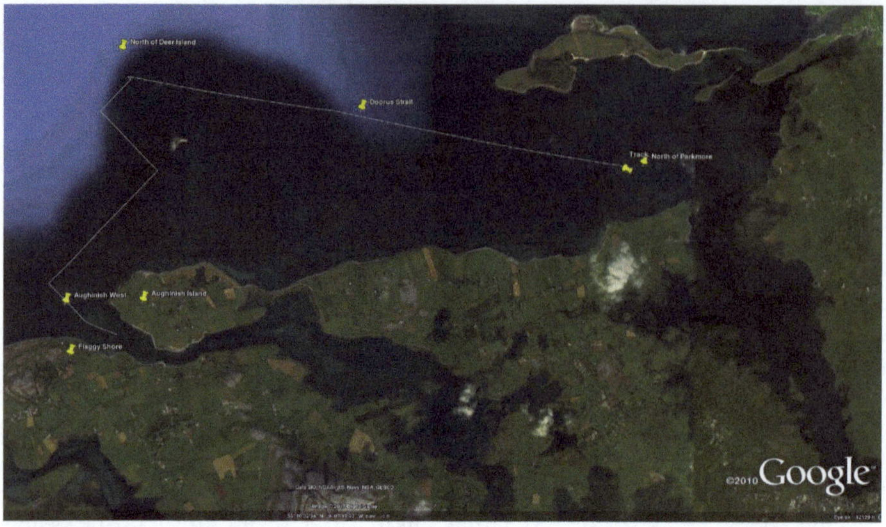

Fig. 4 Simulated boat navigation with VMG calculation

The marks on the water indicate attractors, whereas **Aughinish Island** and **Flaggy Shore** as shown in the figure are repellors. Note that the boat appears to sail over the first waypoint (**Aughinish West**) because the existing tack is considered more favourable, rather than switching to a port tack. On the other hand, the boat does not need to close on **North of Deer Island** because the waypoint has a sufficiently wide radius (0.2 nm in this case). The route to **Doorus Strait** and finally **Parkmore** is relatively direct. As this particular simulation is not using attractors or repellors but simply computing the best VMG for a range of wind angles either side of the bearing to the mark, it is not in a position to avoid obstructions such as *Deer Island* or even *Aughinish*. Conveniently, in this simulation, the boat manages to stay clear of the obvious hazards. Tidal effects, or the loss of GPS signal causing the boat to sail into trouble, is not considered.

In Fig. 5 below, the full complement of attractors and repellors are added to the best-route calculation. The route is almost identical, except for a noticeable arc when the boat is on a reach to the last two waypoints. This is a side-effect of the two repellors, which are causing the boat to try to steer clear of them. The reason for this is because the polar curve for true wind angles in the range of 80°–100° is quite flat. Therefore, choosing any true wind angle in that range will give very similar results for VMG. However, the presence of the two repellors causes the boat to choose the smaller wind angles, and the boat will sail higher initially, until the actual waypoint itself has a strong enough attraction to overcome the repelling effects of **Flaggy Shore** and **Aughinish Island**.

The upwind part of the simulated course shows little difference from the first simulation (without attractors and repellors). This is because more acute upwind angles will cause a dramatic fall-off in predicted hull speed, and bearing away slightly

Fig. 5 Navigation simulation with attractors and repellors

will have a more dramatic impact on the VMG, so the repellors, while important, will not alter the upwind course unless it is hazardous. However, unlike the previous case, the boat will strenuously avoid the two shown hazards.

While various simulations have failed to show any deadlocked scenario, such a possibility could be possible. With sufficient repellors in locations around the boat, and at distances which closely map the polar curve of the boat, it could be possible for all computed headings to show negative VMG values. As the simulated code tends to favour the existing heading unless a new heading offers a marked improvement, and assuming the boat isn't head-to-wind, the boat should sail out of such deadlocked situations, by ignoring the alternatives. If the boat is actually head-to-wind and as a result has a predicted hull speed of 0*knots*, then negative alternatives will not present themselves as more favourable, and the boat will be trapped "in irons". The simulation as written does not check to ensure that the computed heading will produce a reasonable hull speed, but this could be added as part of any future work. In such cases where all computed headings are considered distasteful, the boat could widen the *swing* angle from the normal range of $\pm 45°$ in favour of something wider. Possibly even as wide as 120°. Alternately, the navigation could cease to use the attractor/repellor approach and fall back to an analysis of predicted headings either side of the bearing just to the current waypoint.

Computing the wide range of possible angles for a large list of waypoints is not a trivial exercise, but rough computations show that the compute time is on the order of seconds rather than minutes. Given that the proposed architecture is blended, with a low-level system performing the regular rudder and sail updates while an upper-level system examines the prospective course, this is not an unreasonable performance penalty. It is anticipated that route navigation may occur every 5 or 10 min, depending on the proximity of repellors and attractors to the boat. In the middle of an ocean, hundreds if not thousands of miles from the next waypoint (or hazard), it may be reasonable to recompute the route on a daily basis. In crowded waters with many hazards, the computation may happen continuously, pending fresh updates from the GPS.

7 Future Work

Simulations have considered this navigation algorithm for a wide variety of manufactured situations. However, actual field tests with real tidal set and drift, are necessary to further validate the algorithm under real-world conditions. The tuneable parameters include the amount of tack bias used to dissuade the system from incurring a tack or gybe, the possible state of sailing in terms of the distance to the nearest repellor, and the range of headings to consider in each sailing state. The tack bias can be computed based on real-world data on the detrimental effect of tacking, on the hull speed.

The two key advantages of understanding where the hazards lie, are in the ability to deal with GPS errors or tidal anomalies which can place the boat into a hazardous position, and also to optimize the approach to determining the next waypoint. If the combined distance from the current location to the current waypoint and the distance from the current waypoint to the next waypoint is larger than the distance from the current location to the next waypoint, it is reasonable to assume that the current waypoint has been achieved, and the boat can instead work towards the next waypoint in line. However, this can only work if the boat is able to avoid sailing into hazardous areas such as behind the headland shown earlier. More work is needed in this area.

Future work would also consider the effect of known tidal set and drift as defined in a tidal atlas [20], and add predicted tidal effects to the heading under consideration; choosing a course which sets a small robotic sailboat directly into a strong tidal current cannot be considered ideal.

References

1. AŁstrom, K. J., Hagglund, T. 1995. PID controllers: theory, design, and tuning. S.l.: International Society of Automation.
2. Alves, J., Ramos, T., Cruz, N. 2008. A reconfigurable computing system for an autonomous sailboat. *Journal of the Osterreichische Gesellschaft fur Artificial Intelligence*, pp. 18–24.
3. Alvira, M., Barton, T. 2012. *Small and inexpensive single-board computer for autonomous sailboat control. Robotic sailing*, pp. 105–116.
4. Bowditch, J.I. 2002. *American practical navigator*. S.l.: Defense Mapping Agency Hydrographic Center.
5. Briere, Y. 2006. The first microtransat challenge.
6. Brooks, R. 1986. *A robust layered control system for a mobile robot. Robotics and automation*, pp. 14–23.

7. Burns, R. S. 1995. The use of artificial neural networks for the intelligent optimal control of surface ships. *Journal of Oceanic Engineering*.
8. Donkin, R. 2010. *The future of work*. Springer.
9. Johnson, G. 2008. *Sailing fundamentals*. New York: Simon and Schuster.
10. Langbein, J., Fruhwirth, T. 2011. *A rule-based approach to long-term routing for autonomous sailboats. Robotic sailing*, pp. 195–204.
11. Microtransat, n.d. The Microtransat Challenge. http://www.microtransat.org. Accessed 2017.
12. Minorsky, N. 1922. Directional stability of automatic steered bodies. *Journal of American Society of Naval Engineers*, 280–309.
13. NavSim Technology, n.d. Definitions. http://www.navsim.com/products/help/definitions.html. Accessed May 2017.
14. Raymarine, n.d. Raymarine Autopilot Systems. http://www.raymarine.com/display/?id=581. Accessed May 2017.
15. Rousmaniere, J., and M. Smith. 1999. *The annapolis book of seamanship*. New York: Simon and Schuster.
16. Sauze, C., Neal, M. 2008. Design considerations for sailing robots performing long term autonomous oceanography. In *Proceedings of the International Robotic Sailing Conference*, pp. 21–29.
17. Stelzer, R., Jafarmadar, K. 2007. *A layered system architecture to control an autonomous sailboat. Towards autonomous robotic systems*.
18. Stelzer, R., Jafarmadar, K. 2011. *History and recent developments in robotic sailing. Robotic sailing*, pp. 3–23.
19. Stelzer, R., Proll, T., John, R. I. 2007. Fuzzy logic control system for autonomous sailboats. In *FUZZ-IEEE*, pp. 1–6.
20. United Kingdom Hydrographic Office, n.d. Admiralty Tidal Stream Atlases. https://www.admiralty.co.uk/publications/miscellaneous-tidal-publications/admiralty-tidal-stream-atlases. Accessed May 2017.
21. WRSC, n.d. World Robotic Sailing Championship. http://www.roboticsailing.org/.

Chase, No Straighter: COLREG Compliant Anti-collision Manoeuvres for Crossing Situations

Kjell Ivar Øvergård, Marius Stian Tannum, Per Haavardtun and Jens Kristian Berg Syversen

Abstract The interaction between autonomous vessels and humans is a problem which has not clearly been addressed. One solution for this is that autonomous vessels implement rules and algorithms that mimics the behaviour of trained seafarers that acts in accordance with the International Maritime Collision Regulations (COLREGS). We suggest a simple algorithm that 'chases' the tangent point of the Circle of Rejection that is set outside other vessels. The algorithm utilises the same information that would be used by a human navigator—and it uses this information to guide an autonomous vessel in a way that will satisfy COLREG Rule 8 which requires anti-collision manoeuvres to be positive, timely and with regard to good seamanship.

Keywords COLREGS · Robotic sailing · Anti-collision · Navigation

1 Introduction

Autonomous transport systems are on the verge of becoming a regular part of the everyday life of humans. Currently the automotive industry tests autonomous cars and testing grounds for autonomous vessels have recently been established in the Trondheim Fjord. In the same vein, we see a line of research and development of autonomous sailing vessels which enable robotic sailing (see e.g. [2]). Autonomous sailing vessels have many uses such as surveillance and ocean sampling [8]. Futurists claim that autonomous maritime vessels will be commonplace in the near future [11]. This opening of new technological domain have led to research on the effect of autonomy on the user experience [7] and on the interaction between humans and robots (or other autonomous systems; [17]. However, the transit from human-controlled to autonomous maritime vessels will take time. Hence, autonomous and

K. I. Øvergård (✉) · M. S. Tannum · P. Haavardtun · J. K. B. Syversen
Department of Maritime Operations (KIØ, PH, JBKS), Department of Microsystems (MST), University College of Southeast Norway, Postboks 4, 3199 Borre, Norway
e-mail: koe@usn.no

K. I. Øvergård (ed.), *Robotic Sailing 2017*,
https://doi.org/10.1007/978-3-319-72739-4_7

human-controlled vessels must interact to ensure the safety and efficiency of navigation [11, 17]. Some claim that safety is not the main challenge of maritime navigation when we only focus on the autonomous vessel itself [16] but focusing solely on the autonomous vessel ignores the fact that the vessel's manoeuvres can affect other nearby vessels.

1.1 Anti-collision Regulations

A basic concept of navigation is the ability to follow a pre-planned route in a safe and effective manner and to avoid collisions. In addition, all vessels must abide by the "*International Regulations for Preventing of Collision at Sea*" (COLREGS) [10]. These rules of the road consist of 38 rules which is divided into the following parts general guidelines (Part A), rules regarding sailing and steering and actions by the "stand on" and "give way" vessel (Part B), Lights and shapes (Part C), Sound and light signals (Part D) and exemptions (Part E). The Core aspects of the COLREGS are Rule 13 (Overtaking), 14 (Head-on situations) and 15 (Crossing Situations) [6]. In this paper we only consider Crossing situations.

Rule 13: Overtaking. An overtaking vessel must keep clear of the vessel being overtaken. If a vessel approaches another vessel in the sector of the stern light (22.5 degrees abaft the beam) it is defined to be overtaking the other vessel [9]. During night navigation this is defined by seeing only the stern light and none of the side lights. If a side light is seen then the vessel is seen as crossing (see Rule 15 and 16).

Rule 14: Head-On Situation. When two power-driven vessels are meeting head-on, both in such a way that only the mast lights are visible, then both vessels must alter their course to starboard so that they pass on the port side of the other [9]. The mast light covers a sector +/- 6 degrees centred around the vessel's heading.

Rule 15: Crossing Situation. When two power-driven vessels are crossing (defined as a situation where the vessels see the side light of each other [9]) the vessel which has the other on the starboard side must give way (it is the *Give-Way-vessel* (GWV) [9, 10]. The GWV must avoid crossing in front of the *stand-on-vessel* (SOV). Therefore anti-collision manoeuvres should always be a starboard turn. In addition to rule 13, 14, and 15, there are rules that require that all actions must be done in a positive and timely manner that is in accordance with "good seamanship".

Rule 8: Action to avoid collision. This rule requires that all actions to avoid collisions should be "*positive*", "*made in ample time*" and "*with due regard to the observance of good seamanship*" [9, p. 49]. This means that the course change by the vessel should be large enough to be observed by other vessels. This rule requires that "*any alteration of course and/or speed to avoid collision, must be large enough to be readily apparent to another vessel observing it visually or by radar*" [9, p. 48]. The definition of "readily apparent" is not defined in the regulations, but navigation customs dictate that the GVW should clearly show their intentions by making a turn large enough to show the port side light to the SOV.

Rule 16: Actions by give-way vessel. In addition to Rule 8, Rule 16 also dictates that "*All vessels directed to keep out of the way must take early and substantial action to keep well clear*" [9]. This regulation can be seen as an addition to Rule 8 further explicating the necessity of keeping clear of any conflicting traffic.

Unfortunately, the COLREG's do not specify what a timely and readily apparent manoeuvre is. This means that it is up to the ship master to interpret the situation and make decisions based upon good seamanship, ship sense and the available information at hand to resolve anti-collision situations [14]. In this way, the COLREGS depend on the humans ability to use common sense to exploit the flexibility in the actions prescribed in a rule. Since there are no general rules of the road for USV's the smart move would be to make the vehicles compliant with the COLREGS which is the existing standards for safe navigation [4]. Since each rule is related to an action it is important to translate the COLREGs so the USV can manoeuvre in compliance with the rules of the road. Since the use of COLREGs are dependent upon human meaning-making, it is not suitable for direct input into a vehicle control system [3].

Safe navigation is dependent on the behaviour of other vessels to be predictable. This requires that autonomous vessels behave in the same manner as a well-trained and experienced captain. A fully optimised route with respect to travel distance, time, fuel consumption, safety margins and so on not necessarily lead to a human-like performance.

1.2 Colregs for Autonomous Vessels

Recently Savvaris and colleagues presented an algorithm for collision avoidance for autonomous vessels [15, algorithm 3–6]. Their algorithm gave course changes which where in accordance with COLREGS, however, it is not clear whether the model lead to navigation in accordance with Rule 8—by making a turn large enough to show their port light.

Normally, any uncertainties between two vessels is resolved by radio communication between the two vessels to negotiate a safe and efficient anti-collision manoeuvre. However, automated vessels have a limited ability to communicate and to show intentions other than by manoeuvring. Hence, following an optimal path without showing intentions may create problems for the human in the SOV.

Recently, a defensive approach to Anti-collision manoeuvres was presented by Alves and Cruz [1] where they argue for an approach where a sailboat should always avoid coming in conflict with other vessels (see e.g. Fig. 2 in [1]). Defensive anti-collision manoeuvres means that vessels always manoeuvres to avoid coming in conflict with other vessels. From a safety perspective of the individual vessel this might seem to be a good idea as the sailboat then always would make sure by itself that collisions are avoided—hence, safety at sea is not determined by whether other boats (autonomous or human controlled) also follow COLREGS. However, a defensive approach to anti-collision manoeuvres is very different from the behaviour of other (human-controlled) vessels where it is expected that stand-on-vessels will keep their course and speed in conflict situations. A defensive approach would be unpredictable as the actions do not follow the COLREGS for SOVs. Also, a defensive approach would be inefficient and in opposition to the goal of sailing from A to B. In this regard, the defensive anti-collision manoeuvre may be safe, but surely not efficient. Hence, we need to find middle road between safety (e.g. the defensive manoeuvre [1]) and efficiency (e.g. [15]).

2 Algorithm for "Human" Collision Avoidance

We propose a solution to this challenge that leads to efficient human-like collision avoidance. This should ease the interaction between autonomous vessels and human-controlled vessels.

2.1 Which Rule to Apply

Each of the COLREGS rules 13, 14 and 15 corresponds to a behaviour [3, 5]. To know which rule to implement the COLREGs point towards the light sectors of vessels [9]. Human navigators use these light sectors as navigational aids during visual navigation. The light sectors are defined by the red (port side), green (starboard side) and white (aft and mast) navigation lights. The arc of these lights is given by COLREG 21 [9, 10], and the sectors and the corresponding COLREGS are illustrated below in Fig. 1. Based upon the sectors of the navigation lights presented in Fig. 1 we can determine which rule that applies to a given situation. By subtracting Vessel B's true heading (H_B°) from vessel A's true heading (H_A°) the resulting difference in headings (H_d°) can be used to identify the COLREG that apply to the given situation. So, using the light sectors presented in Fig. 1 and with the equation

$$H_B^\circ - H_A^\circ = H_d^\circ \tag{1}$$

we can identify which H_d° that are associated with the COLREG Rules 13 to 15.

$$IF \quad H_d^\circ \in [-186, -174] \; OR \; [174, 186] \; THEN \quad Rule \; 13 \tag{2}$$

$$IF \quad H_d^\circ \in [-32.5, 32.5] \; THEN \quad Rule \; 14 \tag{3}$$

Rule 15 can have two outcomes—either that our own vessel (V_A) is the GWV or the SOV:

$$IF \quad H_d^\circ \in [-360, -186] \; OR \; [32.5, 174] \; THEN \quad Rule \; 15 \quad (V_A = GWV) \tag{4}$$

$$IF \quad H_d^\circ \in [-174, -32.5] \; OR \; [186, 360] \; THEN \quad Rule \; 15 \quad (V_A = SOV) \tag{5}$$

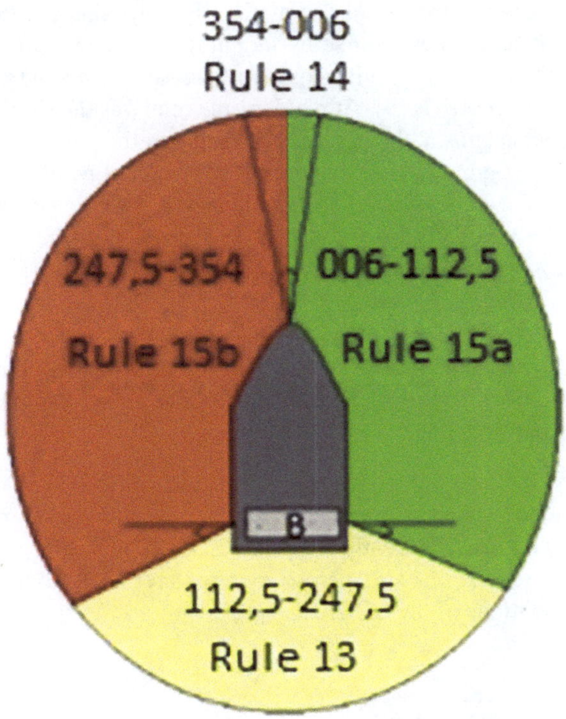

Fig. 1 Sectors based on Navigation lights. In this figure Vessel B has a true heading of 360° and the sectors show the angles of the navigation lights. COLREGs rules are defined according to which sector that other vessels are in relative to Vessel B. Rule 13 refers to an overtaking situation, Rule 14 refers to a head-on situation, Rule 15a to a situation where Vessel A is SOV, Rule 15b to a situation where Vessel B is GWV

By using the light sectors described in COLREG Rule 21 [9, p. 109] we are ensuring that the actions of the autonomous vessels are based upon the same information that would be available to human navigators.

2.2 *The Chasing Strategy for Conflict Situations (Rule 15)*

An important aspect of navigation is the need for positive, timely and readily apparent anti-collision manoeuvres that makes it easy for nearby vessels to observe the change in course and speed visually or by RADAR. We will present a strategy which lead to anti-collision manoeuvre that fulfil the requirements in Rule 8, 15 and 16in the COLREGS [9, 10]. The current chasing strategy is designed and tested in a case of two self-propelled vessels. We have not yet considered sailing vessels nor have we tested for scenarios with more than two vessels (own vessel and one other vessel).

The chasing strategy is based upon the simple ideas that around all other vessels there exists a Circle of Rejection (COR) which defines the area around another vessel which our own vessel should not enter [12]. In Practical navigation the COR corresponds to the concept of Closest Point of Approach (CPA) which is used to make decisions regarding anti-collision manoeuvres. It works in the following way, when an autonomous vessel identify a crossing situation where it is the GWV it will calculate a new course by targeting the tangent point of the COR on the port side of the SOV. This will require a starboard manoeuvre in accordance with Rule 15 in the COLREGs [9]. At regular time intervals the ship's course will be updated so the course aims at the new tangent point on the port side of the SOV's COR. The GWV will continue doing this until it has a clear line of sight to the original Way Point.

The collision avoidance manoeuvre is initiated when the calculation of Time to Closest Point of Approach (TCPA, see [1] for the calculation of TCPA) is less than a specified value. The pseudocode for the chasing algorithm is:

if $CPA < r_{COR}$ **and** $TCPA < T_a$ **then**
  repeat calculate new_{WP} on port side tangent point of SOV's COR
  until new_{WP} is on port side of direct path to old_{WP}
end if

In the pseudocode r_{COR} is the radius of the Circle of Rejection, and T_a is the time set for when you must start an anti-collision manoeuvre. Note that T_a must be decided by humans. Currently the COLREGs [9] do not specify any particular thresholds and the regulations are open so each seafarer needs to adapt to the situation to meet the intentions of the COLREGs [13].

2.3 Calculation of New Course and New Way Points

To calculate the course to the new way point on the tangent point on the port side of the SOV's COR we first calculate the course of vessel A:

$$x_A(t) = V_A \cdot cos\theta_A \cdot t + X_{A_0} \tag{6}$$

$$y_A(t) = V_A \cdot cos\theta_A \cdot t + Y_{A_0} \tag{7}$$

Where the start position of vessel A is (X_{A0}, Y_{A0}), with velocity V_A and heading $\theta_A[0, 2\pi]$. We then calculate the course of vessel B:

$$x_B(t) = V_B \cdot cos\theta_B \cdot t + X_{B_0} \tag{8}$$

$$y_B(t) = V_B \cdot cos\theta_B \cdot t + Y_{B_0} \tag{9}$$

Where the start position of vessel A is (X_{B0}, Y_{B0}), with velocity V_B and heading $\theta_B[0, 2\pi]$. The distance between the vessels $d(t)$ for each time instance is calculated according to the following Pythagorean equation:

$$d(t) = \sqrt{\left(x_A(t) - x_B(t)\right)^2 + \left(y_A(t) - y_B(t)\right)^2} \tag{10}$$

If the measured distance at any point is smaller than the radius of the COR, and the TCPA is smaller than the criterion value the autonomous vessel must evaluate which COLREG rule that applies. If the autonomous vessel finds that it is the GWV it must make a course change towards the tangent point of the COR. Because of Colreg 15, *if the circumstances of the case admit, avoid crossing ahead of the other vessel.* [9, p. 84] the tangent point on the port side of the SOV is chosen. This new way point will always require a starboard manoeuvre—in accordance with COLREGs Rule 15. It will also lead to a course change which is so large that the GWV shows its port lantern to the SOV—thereby adhering to the requirement in Rule 8 that all actions should be "*readily apparent*". Calculating the COR centred at the intruder vessel B $[x_B, y_B]$ with radius r_{COR} is done by the following equation:

$$(x - x_B)^2 + (y - y_B)^2 = r_{COR}^2 \tag{11}$$

The tangent points on the COR around the intruder vessel $B[x_B, y_B]$ seen from the GWV $A[x_A, y_A]$ are calculated with:

$$x_{1,2} = \frac{r_{COR}^2(x_A - x_B) \pm r_{CoR}(y_A - y_B)\sqrt{(x_A - x_b)^2 + (y_A - y_B)^2 - r_{COR^2}}}{(x_A - x_B)^2 + (y_A - y_B)^2} + x_B \tag{12}$$

$$y_{1,2} = \frac{r_{COR}^2(y_A - y_B) \pm r_{COR}(x_A - x_B)\sqrt{(x_A - x_b)^2 + (y_A - y_B)^2 - r_{COR}^2}}{(x_A - x_B)^2 + (y_A - y_B)^2} + y_B \tag{13}$$

The new heading, θ_t to the chosen tangent point $[x, y]$ for GWV is calculated by the equation:

$$\theta_t = tan^{-1}\left(\frac{x_a - x}{y_a - y}\right) \tag{14}$$

Note that this do not calculate collision risk, but rather calculate the new course of the SOV [15].

3 Results and Discussion

Simulations of the algorithms with two self-propelled vessels was done in Scilab© version 5.5.2. Result of the simulations are presented in Figs. 2 and 3 which shows that the distance between the GWV and the SOV during this simulation always exceeds the radius of the COR. We can also see that the manoeuvre is large and that the GWV makes a turn that allows the SOV to see the GWV's port side light—thereby indicating clearly to the SOV that the GWV have made a large manoeuvre to starboard that effectively avoids collision. Of course, such a solution is not efficient in the sense that it takes a longer route than necessary.

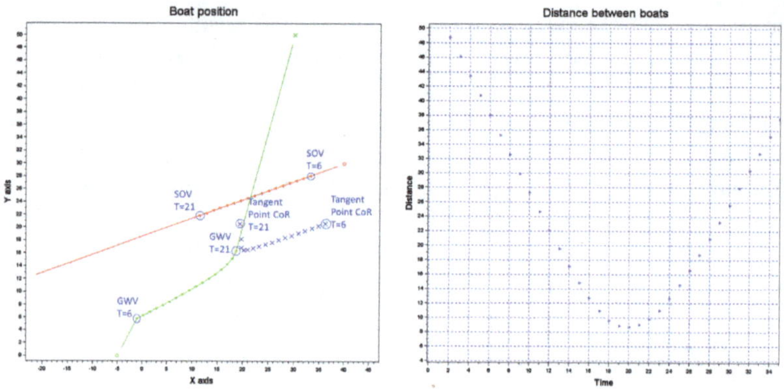

Fig. 2 Simulations of the chasing strategy, *Left Side*: Route sailed: The position of the SOV is marked as a red straight line and the GWV's position as a green line. The starting point of both vessels are marked with O for the simulation (time = 0). TCPA parameter is set to 15 time units and the COR radius set to 8. The SOV's heading is 35° The GWV's heading is 254° and its first starboard course change occurs at $T = 6$ ($X = -0.95$, $Y = 5.79$) aiming at the tangent point ($X = 36.17$, $Y = 20.63$) on the COR for the SOV currently at position ($X = 33.2$, $Y = 28.1$). The following tangent points are marked blue and represents new headings for the GWV. A new course is calculated at each new time step until $T = 21$ when the GWV has clear sight to the final way point at point ($X = 30$, $Y = 50$) when it chooses a course towards the original way point instead of the last tangent point. *Right Side* shows the distances between the two vessels over time

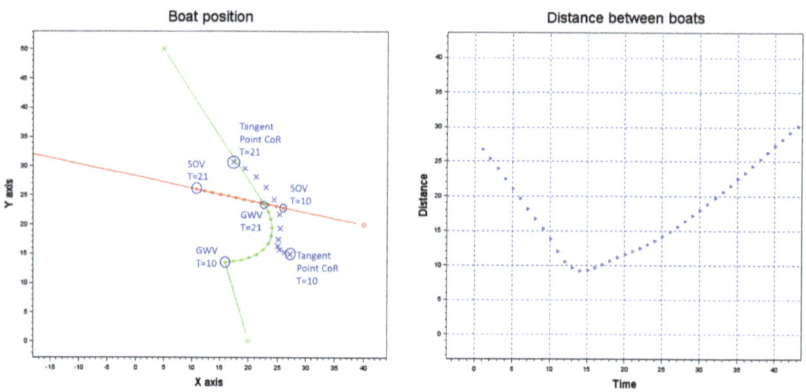

Fig. 3 Simulations of the chasing strategy, *Left Side*: Route sailed: The position of the SOV is marked as a red straight line and the GWV's position as a green line. The starting point of both vessels are marked with O for the simulation (time = 0). TCPA parameter is set to 10 time units and the COR radius set to 8. The SOV's heading is 282° The GWV's heading is 343° and its first starboard course change occurs at $T = 10$ ($X = 15.94$, $Y = 13.55$) aiming at the tangent point ($X = 27.14$, $Y = 14.49$) on the COR for the SOV currently at position ($X = 26.15$, $Y = 22.88$). The following tangent points are marked blue and represents new headings for the GWV. A new course is calculated at each new time step until $T = 21$ where the GWV has clear line of sight to the final way point ($X = 5$, $Y = 50$) when it chooses a course towards the original way point instead of the last tangent point. *Right Side* shows the distances between the two vessels over time

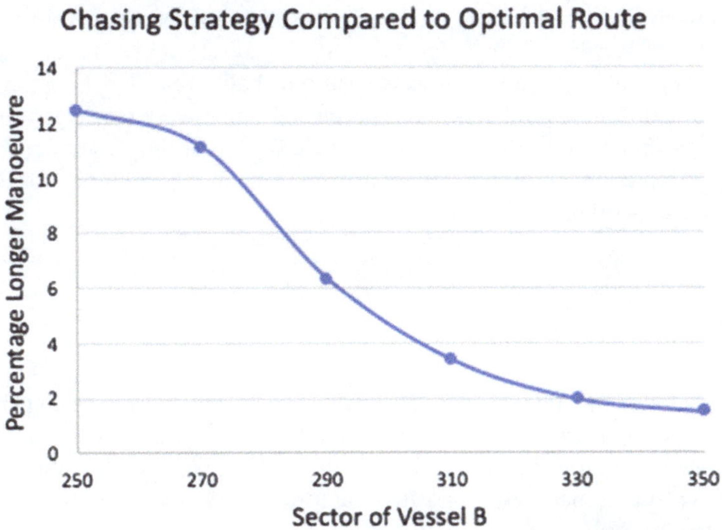

Fig. 4 How much longer the distance sailed when following the chasing strategy as compared to an optimal strategy. The longer distance is dependent upon the angle in which the GWV approaches the SOV

The chasing strategy leads to manoeuvring in accordance with COLREGs, but still it is less efficient than an optimal route with respect to length. The optimal route makes only one course change that aims at the tangent point the aft of the SOVs COR and then one course change when it is directly aft of the other vessel. Given that both vessels sail at the same speed, simulations indicate that the chasing strategy at maximum leads to a route which is not more than 13% longer. This happens when the GWV comes in towards the SOV at an angle of 250° relative to the SOVs heading. See Fig. 4 for an overview of how much longer the GWV must navigate as a function of the angle in which it approaches the SOV. The loss in efficiency is countered by the clear and positive anti-collision manoeuvres and that the vessels are also able to stay away from other vessel's CoR. Of course, the added length of the avoidance manoeuvre relative to an optimal manoeuvre is dependent on the relative speed of the vessels, hence real losses of efficiency might deviate somewhat from the result in Fig. 4.

3.1 Limitations

A drawback with the chasing strategy is the need to make many small course changes after the large initial course change. From a navigation perspective this is an inelegant and cumbersome strategy as humans are not good at monitoring small heading changes.

Also, our simulations have only focused on a self-propelled vessel (not a sailboat) in a navigation scenario with only one other vessel. However, the chasing strategy is still applicable for sailing vessels—but the challenges of making the chasing manoeuvre in different wind conditions needs additional work (see e.g. [1])

Future research will test the viability of our chasing algorithm in settings with poor weather, multiple other crossing vessels as well as involving solutions for Head-On Situations and Overtaking situations.

4 Conclusion

We have presented an algorithm for an anti-collision strategy in crossing situations that uses the same information as humans would have used (navigation light sectors) and that acts in accordance with COLREGs Rules 8, 15 and 16. The algorithm will allow the vessel to make large, positive, and timely manoeuvres that are readily apparent for the SOV.

References

1. Alves, J.C., and N.A. Cruz. 2016. Ais-enabled collision avoidance strategies for autonomous sailboats. In *Robotic sailing 2015*, 77–87. Springer Science & Business Media.
2. Alves, J.C., and N.A. Cruz. 2016. *Robotic sailing 2016: Proceedings of the 9th international robotic sailing conference*, Springer Science & Business Media.
3. Benjamin, M., J. Curcio, J. Leonard, and P. Newman. 2006. Protocol-based colregs collision avoidance navigation between unmanned marine surface craft. *Journal of Field Robotics* 23:5.
4. Benjamin, M.R., and J.A. Curcio. 2004. Colregs-based navigation of autonomous marine vehicles. In *Autonomous underwater vehicles, 2004 IEEE/OES* 32–39. IEEE.
5. Benjamin, M.R., J.A. Curcio, J.J. Leonard, and P.M. Newman. 2006. Navigation of unmanned marine vehicles in accordance with the rules of the road. In *Robotics and automation, 2006. ICRA 2006. Proceedings 2006 IEEE international conference*, 3581–3587, IEEE.
6. Campbell, S., and W. Naeem. 2012. A rule-based heuristic method for colregs-compliant collision avoidance for an unmanned surface vehicle. *IFAC Proceedings Volumes* 45(27): 386–391.
7. Christ, P.F., F. Lachner, A. Hösl, B. Menze, K. Diepold, and A. Butz. 2016. Human-drone-interaction: A case study to investigate the relation between autonomy and user experience. In *European conference on computer vision*, 238–253, Springer.
8. Cruz, N.A., and J.C. Alves. 2008. Autonomous sailboats: An emerging technology for ocean sampling and surveillance. In *OCEANS 2008*, 1–6. IEEE.
9. van Dokkum, K. 2007. A fully illustrated textbook: The COLREGS Guide. DOKMAR.
10. IMO. 1972. Convention on the International Regulations for preventing collision at sea (COLREGs). International Maritime Organisation.
11. Levander, O. 2017. Autonomous ships on the high seas. *IEEE Spectrum* 54 (2): 26–31.
12. Naeem, W., and G.W. Irwin. 2011. Evasive decision making in uninhabited maritime vehicles. *IFAC Proceedings Volumes* 44 (1): 12833–12838.
13. Øvergård, K.I., S. Nazir, and A.S. Solberg. 2017. Towards automated performance assessment for maritime navigation. *TransNav, the International Journal on Marine Navigation and Safety of Sea Transportation* 11 (2): 43–48. https://doi.org/10.12716/1001.11.02.03.

14. Porathe, T., J. Prison, and Y. Man. 2014. Situation awareness in remote control centres for unmanned ships. In *Proceedings of human factors in ship design & operation*, p. 93, Feb 26–27, London, UK.
15. Savvaris, A., H. Niu, H. Oh, and A. Tsourdos. 2014. Development of collision avoidance algorithms for the c-enduro usv. In *Proceedings of the 19th IFAC world congress (IFAC 2014)*, vol. 47, 12174–12181.
16. Stenman, A., and S. Öhland. 2017. *Interaction between unmanned vessels and colregs*. Yrkeshögskolan Novia: Tech. rep.
17. Wolf, I. 2016. The interaction between humans and autonomous agents. In Autonomous driving, 103–124. Springer.

Kolibri 360 GmbH
Ferdinand-Jühlke-Straße 7
99095 Erfurt, Deutschland
produktsicherheit@kolibri360.de